Springer Series in Reliability Engineering

Series Editor

Hoang Pham, Industrial and Systems Engineering, Rutgers, The State University of New Jersey, Piscataway, NJ, USA

Today's modern systems have become increasingly complex to design and build, while the demand for reliability and cost effective development continues. Reliability is one of the most important attributes in all these systems, including aerospace applications, real-time control, medical applications, defense systems, human decision-making, and home-security products. Growing international competition has increased the need for all designers, managers, practitioners, scientists, and engineers to ensure a level of reliability of their product before release at the lowest cost. The interest in reliability has been growing in recent years and this trend will continue during the next decade and beyond.

The Springer Series in Reliability Engineering publishes books, monographs, and edited volumes in important areas of current theoretical research development in reliability and in areas that attempt to bridge the gap between theory and application in areas of interest to practitioners in industry, laboratories, business, and government.

Now with 100 volumes!

Indexed in Scopus and EI Compendex

Interested authors should contact the series editor, Hoang Pham, Department of Industrial and Systems Engineering, Rutgers University, Piscataway, NJ 08854, USA. Email: **hopham@soe.rutgers.edu, or Anthony Doyle, Executive Editor, Springer, London. Email: anthony.doyle@springer.com.**

Yoshinobu Tamura · Shigeru Yamada

Applied OSS Reliability Assessment Modeling, AI and Tools

Mathematics and AI for OSS Reliability Assessment

Yoshinobu Tamura
Graduate School of Sciences
and Technology for Innovation
Yamaguchi University
Ube, Japan

Shigeru Yamada
Tottori University
Tottori, Japan

ISSN 1614-7839　　　　　　　ISSN 2196-999X　(electronic)
Springer Series in Reliability Engineering
ISBN 978-3-031-64802-1　　　ISBN 978-3-031-64803-8　(eBook)
https://doi.org/10.1007/978-3-031-64803-8

© The Editor(s) (if applicable) and The Author(s), under exclusive license to Springer Nature Switzerland AG 2024

This work is subject to copyright. All rights are solely and exclusively licensed by the Publisher, whether the whole or part of the material is concerned, specifically the rights of translation, reprinting, reuse of illustrations, recitation, broadcasting, reproduction on microfilms or in any other physical way, and transmission or information storage and retrieval, electronic adaptation, computer software, or by similar or dissimilar methodology now known or hereafter developed.
The use of general descriptive names, registered names, trademarks, service marks, etc. in this publication does not imply, even in the absence of a specific statement, that such names are exempt from the relevant protective laws and regulations and therefore free for general use.
The publisher, the authors and the editors are safe to assume that the advice and information in this book are believed to be true and accurate at the date of publication. Neither the publisher nor the authors or the editors give a warranty, expressed or implied, with respect to the material contained herein or for any errors or omissions that may have been made. The publisher remains neutral with regard to jurisdictional claims in published maps and institutional affiliations.

This Springer imprint is published by the registered company Springer Nature Switzerland AG
The registered company address is: Gewerbestrasse 11, 6330 Cham, Switzerland

If disposing of this product, please recycle the paper.

*This work is dedicated to my mother,
Chizuko Tamura passed away on July, 2023*

—Yoshinobu Tamura

Preface

We have been energetically proposing several reliability assessment methods for open source software for last two decades. Its measurement and management technologies for open source software are essential to produce and maintain quality/reliable system by using open source software. This book will serve as a textbook and reference book for graduate students and researchers in reliability for open source software, modeling, and deep learning. Several latest methods of reliability assessment for open source software are introduced. We aim to present the state-of-the-art of open source software reliability measurement and assessment based on the stochastic modeling and deep learning approaches. For example, stochastic differential equation models, two dimensional stochastic differential equation models, deep learning methods, tools are presented.

Chapter 1 introduces several aspects of software reliability, open source software development paradigm, and its applications. Also, the basic concept of mathematical modeling based on probability theory and statistics required to describe the software fault-detection phenomena or the software failure-occurrence phenomena and estimate the software reliability quantitatively is introduced.

Chapter 2 describe the basic concept of stochastic differential equation modeling. Chapter 3 introduces two dimensional stochastic differential equation models. Chapter 4 is the jump diffusion process model as an application of the stochastic differential equation modeling for OSS reliability analysis.

Chapters 5 and 6 describe the methods of cyclically two dimensional stochastic differential equation modeling and cyclically two dimensional jump diffusion process modeling considering the edge OSS computing environment. Chapter 7 shows several numerical examples based on the developed software tool.

Chapters 8, 9 and 10 provide several deep learning approaches by using the OSS fault big data sets. In particular, the Wiener process and jump diffusion process are used for data preprocessing as a new approach. Moreover, several numerical examples for the proposed methods of OSS reliability assessment are based on several actual data sets recorded in the bug tracking systems for open source projects. The proposed method will be helpful to assess the OSS reliability. Chapter 11 shows that the tool will be useful for software managers and engineers to assess the reliability of open source software by applying the deep learning.

This book will serve as a textbook and reference book for graduate students and researchers in OSS reliability modeling with application of project management factor and fault big data treatment. Several stochastic reliability analyses of OSS computing with application are introduced along with actual OSS project data.

We would like to express our sincere appreciation to Professor Hoang Pham, Rutgers University, and editor Anthony Doyle, Springer–Verlag, London, for providing the opportunity for us to author this book.

Tottori, Japan Shigeru Yamada
Yamaguchi, Japan Yoshinobu Tamura
December 2023

Contents

1 Open Source Software Reliability 1
   1.1   Introduction 1
   1.2   Definitions 4
   1.3   Software Reliability Measurement and Assessment 5
   1.4   Open Source Software Reliability 7
        1.4.1   Brief Summary of Open Source Software 7
        1.4.2   Development Paradigm of OSS 8
   References 9

2 Stochastic Differential Equation Model for OSS Reliability Analysis 11
   2.1   Introduction 11
   2.2   Software Reliability Modeling for Big Fault Data 12
   2.3   Reliability Assessment Measures 13
   2.4   Cost Optimization Based on the Stochastic Differential Equation Model 13
   2.5   Numerical Examples 14
   2.6   Concluding Remarks 27
   References 29

3 Two Dimensional Stochastic Differential Equation Model for OSS Reliability Analysis 31
   3.1   Introduction 31
   3.2   Two Dimensional Wiener Process Modeling 32
   3.3   Cost Optimization Based on Two Dimensional Stochastic Differential Equation Model 34
   3.4   Numerical Examples 35
   3.5   Concluding Remarks 36
   References 37

4	**Jump Diffusion Process Model for OSS Reliability Analysis**	39
	4.1 Introduction	39
	4.2 Jump Process Modeling	39
	4.3 Cost Optimization Based on Jump Diffusion Process Model	41
	4.4 Numerical Examples	42
	4.5 Concluding Remarks	43
	References	44
5	**Cyclically Two Dimensional Stochastic Differential Equation Modeling**	45
	5.1 Introduction	45
	5.2 OSS Reliability Assessment Modeling with Two Dimensional Cyclic Wiener Process	46
	5.3 Cost Optimization Based on Cyclically Two Dimensional Stochastic Differential Equation Model	51
	5.4 Numerical Examples	51
	5.5 Concluding Remarks	58
	References	58
6	**Cyclically Two Dimensional Jump Diffusion Process Modeling**	61
	6.1 Introduction	61
	6.2 Weighted Jump Diffusion Process Modeling	61
	6.3 Cost Optimization Based on Cyclically Two Dimensional Jump Diffusion Process Model	63
	6.4 Numerical Examples	64
	6.5 Concluding Remarks	65
	References	66
7	**Three Dimensional Tool Based on Noisy Model**	69
	7.1 Development of Prototype Tool	69
	7.2 Performance Illustrations of the Developed 3D Application	71
	7.2.1 Data Set for Edge OSS Computing	71
	7.2.2 Performance Results	74
	7.3 Concluding Remarks	75
	References	78
8	**Deep Learning Method Based on Fault Big Data Analysis for OSS Reliability Assessment**	81
	8.1 Introduction	81
	8.2 Big Fault Data Analysis by Using Neural Network	82
	8.3 Big Fault Data Analysis by Using Deep Learning	83
	8.4 Numerical Examples	84
	8.5 Concluding Remarks	146
	References	149

9 Deep Learning Approach for OSS Reliability Assessment Considering Wiener Process ... 151
9.1 Introduction ... 151
9.2 Wiener Deep Learning Approach ... 152
9.3 Numerical Examples ... 153
9.4 Concluding Remarks ... 162
References ... 162

10 Deep Learning Approach for OSS Reliability Assessment Considering Jump Diffusion Process ... 165
10.1 Introduction ... 165
10.2 OSS Reliability Assessment Based on Deep Learning Model with Jump Diffusion Process ... 166
10.3 Numerical Examples ... 167
10.4 Concluding Remarks ... 176
References ... 176

11 Performance Illustrations of the Developed Application Tool Based on Deep Learning ... 179
11.1 Data Set for Edge OSS Computing ... 179
11.2 Estimation Results ... 179
11.3 Concluding Remarks ... 184
Reference ... 184

12 Exercise ... 185
12.1 Exercise 1 ... 185
12.2 Exercise 2 ... 185
12.3 Exercise 3 ... 187
12.4 Exercise 4 ... 188

Open Source Software Reliability

1.1 Introduction

In recent years, many computer system failures have been caused by software faults which were introduced during the software development process. This is an inevitable problem since a software system installed in the computer system is an intellectual product consisting of documents and source programs developed by human activities. Then, Total Quality Management (TQM) is considered to be one of the key technologies to produce more highly reliable software products. In case of TQM for software management, all phase of the development process, i.e. specification, design, coding, and testing, have to be controlled systematically to prevent software fault-introduction as much as possible and to detect the introduced faults in the software system as early as possible. Basically, the concept of TQM means to assure the quality of the intermediate products in each phase to the next phase. Particularly, quality control executed at the testing phase which is the last stage of the software development process is very important. During the testing phase, the product quality and the software performance during the operation phase are evaluated and assured. Concretely, a lot of software faults introduced in the software system through the first three phases of the development process by human activities are detected, corrected, and removed. It is a general software development process so-called a water-fall diagram in Fig. 1.1.

Under a software development process based on TQM, it is necessary to define and measure the software quality. Generally, software quality realized through the development process can be distinguished between product quality and process quality. The former is the attributes of the products resulting from the software development. This includes the clarity of the specification documents, the design documents, and the source code, the traceability, the reliability, and the test coverage. The latter is the attributes contributed by the software development environment. This includes the rigor of the production technology, the productivity of the development

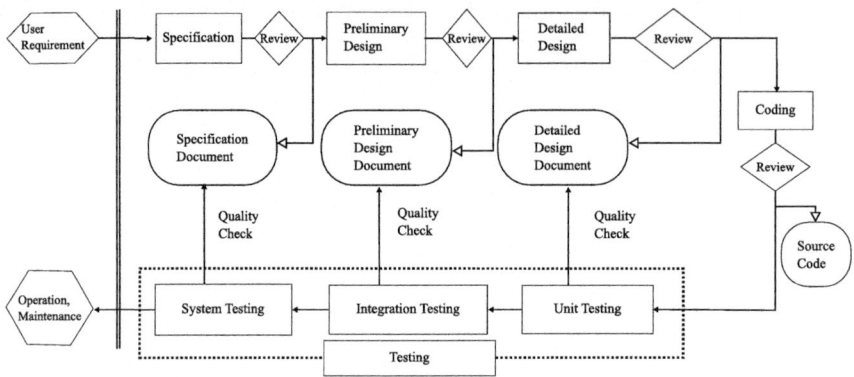

Fig. 1.1 A general software development process (water-fall paradigm)

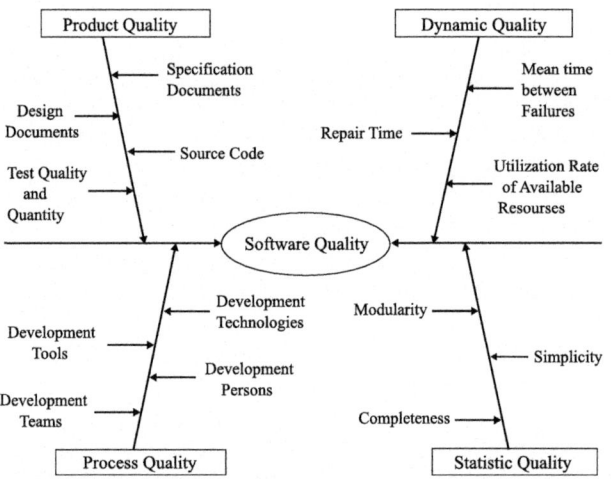

Fig. 1.2 The elements of software quality based on a cause-and-effect diagram

tools, the ability or skill of the development persons, the communicativeness among the development team, and the availability of the development facilities. From the measurement point of view, software quality can be distinguished between dynamic quality and static quality. The former is the attributes confirmed by executing the program on the computers and examining the dynamic behavior. The latter is the attributes validated by reviews and inspections at each phase without executing the program. These elements of software quality are summarized by a cause-and-effect diagram in Fig. 1.2.

1.1 Introduction

Recently, recognizing the importance of standardizing quality characteristics, ISO (International Standard Organization) has studied and summarized six quality characteristics: functionality, reliability, usability, efficiency, maintainability, and portability (ISO/IEC 9126). Particularly, the characteristic of reliability is important as "must-be quality." From the view points of product quality and dynamic quality above, *software reliability* is defined as the probability that a software will not cause the failure of a system for a specified time under specified conditions. In particular, the diagram of software reliability modeling is shown in Fig. 1.3.

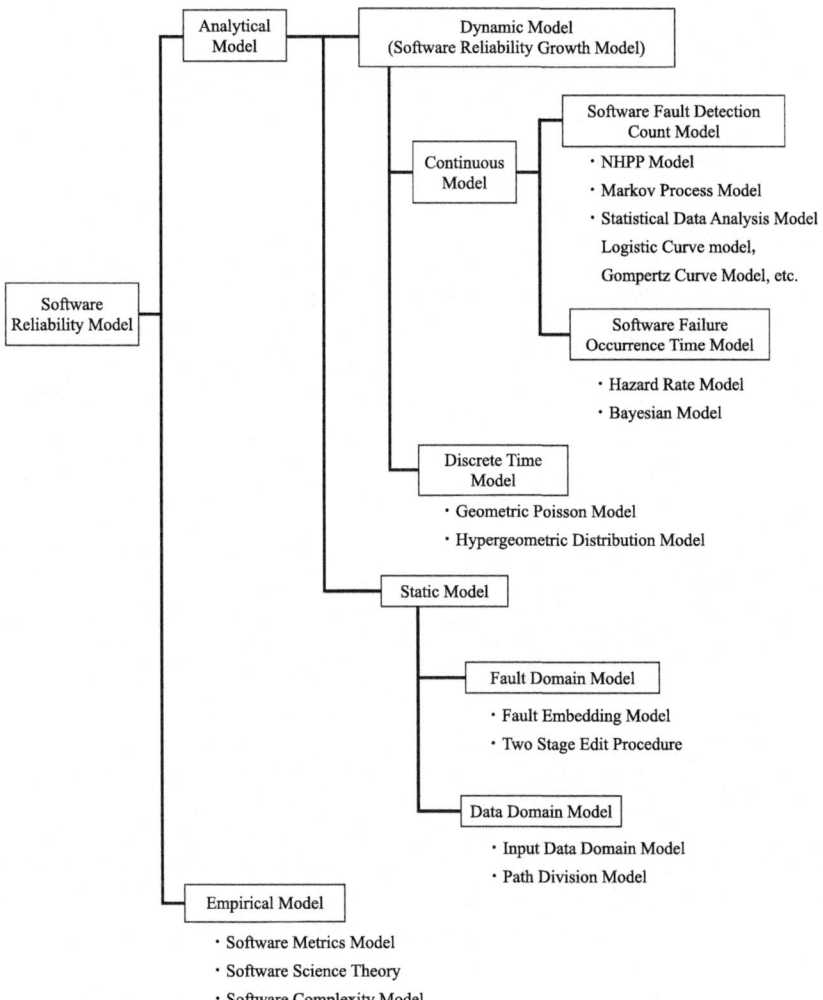

Fig. 1.3 The diagram of software reliability modeling

1.2 Definitions

We compare the characteristics of software reliability with those of hardware reliability as follows:

Software Reliability

1. Software failures can be due to no wear-out phenomenon.
2. Software reliability is, inherently, determined during the earlier phase of the development process, i.e., specification and design phases.
3. Software reliability cannot be improved by redundancy with identical versions.
4. A verification method of software reliability has not been established.
5. A maintenance technology is not established since the market of software products is rather recent.

Hardware Reliability

1. Hardware failures can be due to wear.
2. Hardware reliability is affected by deficiencies injected during all phases of the development, operation, and maintenance.
3. Hardware reliability can be improved by redundancy with identical units.
4. A testing method of hardware reliability is established and standardized.
5. A maintenance technology is advanced since the market of hardware products is established and the user environment is seized.

Generally, a software failure caused by software faults latent in the software system cannot occur expected for a special occasion when a set of special data is put into the system under a special condition, i.e. the program path including software faults is executed. Therefore, the software reliability is determined by the input data and the internal condition of the program. We summarize the definitions of the terms related to the software reliability in the following.

A *software system* is a product which consists of the programs and documents as a result of the software development process discussed in previous section. Specification derived by analyzing user requirements for the software system is a document which describes the expected performance of the system. When the software performance deviates from the specification and the output variable has an improper value or the normal processing is interrupted, it is said that a software failure occurs. That is, *software failure* is defied as an unacceptable departure of program operation from the program requirements. The cause of software failure is called a software fault. Then, *software fault* is defined as a defect in the program which causes a software failure. The software fault is usually called software bug. *Software error* is defined as human action that results in the software system containing a software fault. Thus, the software fault is considered to be a manifestation of software errors.

1.3 Software Reliability Measurement and Assessment

Generally, a mathematical model based on probability theory and statistics is useful to describe the software fault-detection phenomena or the software failure-occurrence phenomena and estimate the software reliability quantitatively. During testing phase in the software development process, software faults are detected and removed with a lot of test-effort expenditures. Then, the number of remaining faults in the software system is decreasing as the testing goes on. This means that the probability of software failure-occurrence is decreasing so the software reliability is increasing and the time-interval between software failures becomes longer with the testing time. A mathematical tool which treats such software reliability aspect is a *software reliability growth model* [1].

Based on the definitions discussed previous section, we can make a software reliability growth model based on the assumptions for actual environments during the testing phase or the operation phase. Then, we can define the following random variables on the number of detected faults and the software failure-occurrence time.

$N(t)$: the cumulative number of detected software faults
(or the cumulative number of observed software failures) up to testing time t,
S_i : the i-th software failure occurrence time ($i = 1, 2, \ldots$; $S_0 = 0$),
X_i : the time-interval between $(i-1)$-st and i-th software failures ($i = 1, 2, \ldots$; $X_0 = 0$).

Figure 1.4 shows occurrence of event $\{N(t) = i\}$ since i faults have been detected up to time t. From these definitions, we have

$$S_i = \sum_{k=1}^{i} X_k, \qquad X_i = S_i - S_{i-1}. \tag{1.1}$$

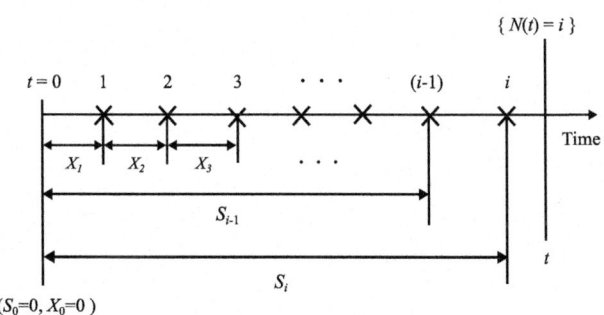

(X: Software fault detection or software failure occurrence)

Fig. 1.4 The stochastic quantities related to a software fault-detection phenomenon or a software failure occurrence phenomenon

Assuming that the hazard rate, i.e. the *software failure rate*, for X_i ($i = 1, 2, \ldots$), $z_i(x)$, is proportional to the current number of residual faults remaining in the system, we have

$$z_i(x) = (N - i + 1)\lambda(x), \quad i = 1, 2, \ldots, N; \ x \geq 0, \ \lambda(x) > 0, \quad (1.2)$$

where N is the initial fault content and $\lambda(x)$ the software failure rate per fault remaining in the system at time x. If we consider two special cases in Eq. (1.2) as

$$\lambda(x) = \phi, \quad \phi > 0, \quad (1.3)$$
$$\lambda(x) = \phi x^{m-1}, \quad \phi > 0, m > 0, \quad (1.4)$$

then two typical software hazard rate models, respectively called the Jelinski-Moranda model and the Wagoner model can be derived, where ϕ and m are constant parameters. Usually, it is completely fault free or failure free. Then, we have a software hazard rate model called the Moranda model for the case of the infinite number of software failure occurrences as

$$z_i(x) = Dk^{i-1}, \quad i = 1, 2, \ldots; \ D > 0, \ 0 < k < 1, \quad (1.5)$$

where D is the initial software hazard rate and k the decreasing ratio, Eq. (1.5) describes a software failure-occurrence phenomenon where a software system has high frequency of software failure occurrence during the early stage of the testing or the operation phase and it gradually decreases thereafter, Based on the software hazard rate models above, we can derive several software reliability assessment measures. For example, the software reliability function for X_i ($i = 1, 2, \ldots$) is given as

$$R_i(x) = \exp\left[-\int_0^x z_i(x) dx\right], \quad i = 1, 2, \ldots. \quad (1.6)$$

Further, we also discuss NHPP models, which are modeled for random variable $N(t)$ as typical SRGM's, In the NHPP models, a nonhomogeneous Poisson process (NHPP) is assumed for the random variable $N(t)$, the distribution function of which is given by

$$\Pr\{N(t) = n\} = \frac{H(t)^n}{n!} \exp[-H(t)], \quad n = 1, 2, \ldots,$$
$$H(t) \equiv E[N(t)] = \int_0^t h(x) dx, \quad (1.7)$$

where $\Pr[\cdot]$ and $E[\cdot]$ mean the probability and expectation, respectively. $H(t)$ in Eq. (1.7) is called a mean value function which indicates the expectation of $N(t)$, i.e., the expected cumulative number of faults detected (or the expected cumulative number of software failures occurred) in the time-interval $(0, t]$, and $h(t)$ in Eq. (1.7) called an intensity function which indicated the instantaneous fault-detection rate at time t. From Eq. (1.7), various software reliability assessment measures can be derived. For examples, the expected number of faults remaining in the system at time t is given by

$$n(t) = a - H(t), \quad (1.8)$$

1.4 Open Source Software Reliability

where $a \equiv H(\infty)$, i.e., parameter a denotes the expected initial fault content in the software system. Given that the testing or the operation has been going on up to time t, the probability that a software failure does not occur in the time-interval $(t, t + x](x \geq 0)$ is given by conditional probability $\Pr\{X_i > x | S_{i-1} = t\}$ as

$$R(x|t) = \exp[H(t) - H(x + t)], \qquad t \geq 0, \ x \geq 0. \tag{1.9}$$

$R(x|t)$ in Eq. (1.9) is a so-called software reliability. Measures of MTBF (mean time between software failures or fault-detections) can be obtained follows:

$$MTBF_l(t) = \frac{1}{h(t)}, \tag{1.10}$$

$$MTBF_c(t) = \frac{t}{H(t)}. \tag{1.11}$$

MTBFs in Eqs. (1.10) and (1.11) are called instantaneous MTBF and cumulative MTBF, respectively. It is obvious that the lower the value of $n(t)$ in Eq. (1.8), the higher the value $R(x|t)$ for specified x in Eq. (1.9), or the longer the value of MTBFs in Eqs. (1.10) and (1.11), the higher the achieved software reliability is. Then, analyzing actual test data with accepted NHPP models, there measures can be utilized to assess software reliability during the testing or operation phase, where statistical inferences, i.e. parameter estimation and goodness-of-fit test, are usually performed by a method of maximum-likelihood.

To assess the software reliability actually, it is necessary to specify the mean value function $H(t)$ in Eq. (1.7). Many NHPP models considering the various testing or operation environments for software reliability assessment have been proposed in the last decade. As discussed above, a software reliability growth is described as the relationship between the elapsed testing or operation time and the cumulative number of detected faults and can be shown as the reliability growth curve mathematically.

Among the NHPP models, exponential and modified exponential SRGM's are appropriate when the observed reliability growth curve shows an exponential curve. Similarly, delayed S-shaped and inflection S-shaped SRGM's are appropriate when the reliability growth curve is S-shaped.

1.4 Open Source Software Reliability

1.4.1 Brief Summary of Open Source Software

All over the world people can obtain the information at the same time by growing rate of Internet access around the world in recent years. In accordance with cloud computing, it is increasing public awareness of the importance of online real-time and interactive functions. Therefore, the distributed software development paradigm such as cloud computing are expanding at an explosive pace. Especially, new development paradigm such as edge computing by using 5G network technologies have been in heavy usage by the software developers. The open source project contains special features so-called edge computing is developed in all parts of the world. The successful experience of adopting the distributed development model is open source projects. Especially, OSS (open source software) is frequently applied as server use, instead of client use. Then, such an open source system development is still ever-expanding now.

A computer-software is developed by human work, therefore many software faults must be introduced into the software product during the development process. Thus, these software faults often cause complicated break-downs of computer systems. Recently, it becomes more difficult for the developers to produce highly-reliable software systems efficiently because of the diversified and complicated software requirements. Therefore, it is necessary to control the software development process in terms of quality and reliability.

Basically, software reliability can be evaluated by the number of detected faults or the software failure-occurrence time. Especially, *software reliability models* which can describe software fault-detection or failure-occurrence phenomena are called SRGM's [1]. The SRGM's are useful to assess the reliability for quality control and testing-process control of software development such as Fig. 1.3. In the past, our research group has proposed several reliability assessment methods based on SRGM's for OSS [2].

1.4.2 Development Paradigm of OSS

It is difficult for software managers to assess OSS reliability because of the differences among the development style of OSS and traditional software. Therefore, it is important to assess and manage considering the characteristics of the OSS development paradigm such as Fig. 1.5.

In particular, OSS have several versions of the development process as follows:

1. Bug Fix Version (most urgent issue such as patch)
2. Minor Version (minor revision by the addition of a component and module)
3. Major Version (significant revision for specification)

Also, the version number of OSS is generally described as the "(Major version number. Minor version number. Revision number. Build number)", e.g., (3.2.2101.1107). In this case, the major version number is 3, minor version number is 2, revision number is 2101, and build number is 1107. There are several versions for each OSS.

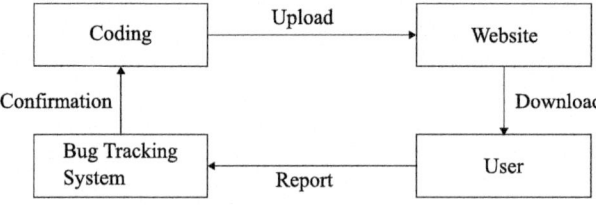

Fig. 1.5 The development cycle of open source project

Therefore, it is known that it is difficult for software managers to select the appropriate OSS, because several OSS are uploaded to the website of open source project.

This book focus on the reliability of OSS with the above mentioned characteristics. Several methods of reliability assessment for OSS are presented in this book. Also, several numerical examples for each method of reliability assessment are given by using actual fault data of OSS. These method introduced in this book may be useful for software managers to assess the reliability of a software system developed using the OSS paradigm.

References

1. Yamada S (2014) Software reliability modeling: fundamentals and applications. Springer, Tokyo/Heidelberg
2. Yamada S, Tamura Y (2016) OSS reliability measurement and assessment. Springer series in reliability engineering. Springer

Stochastic Differential Equation Model for OSS Reliability Analysis

2.1 Introduction

Many OSS's have been embedded in the software development paradigm. For example, the software development in several application software, mobile software, and cloud computing use the OSS, because of the cost reduction, quick delivery, and work saving. In particular, the cloud OSS has a unique characteristic such as the provisioning processes, the network-based operating, and the big data. Therefore, it is important for the cloud managers to assess the reliability in terms of the big data factor during the operation phase of cloud computing. At present, the amount of the data registered in the bug tracking system become large by using the cloud data operation.

Historically, *software reliability models* which describe software fault-detection or failure-occurrence phenomena in the system testing-phase are called SRGM's [1]. The SRGM's are useful to assess the reliability quantitatively for quality control and testing-process control of software development. Many SRGM's have been applied to assess the reliability for quality management and testing-progress control of software development [1–4].

Especially, our research group has proposed the effort optimization method based on many effort prediction models for the OSS development effort data [5–12]. On the other hand, we focus on the OSS development fault data set without effort data one in this chapter.

In this chapter, we propose a simple stochastic differential equation model for the OSS fault data sets. Also, we show several numerical examples of several software reliability assessment measures derived from our model for the big fault data.

© The Author(s), under exclusive license to Springer Nature Switzerland AG 2024
Y. Tamura and S. Yamada, *Applied OSS Reliability Assessment Modeling, AI and Tools*, Springer Series in Reliability Engineering,
https://doi.org/10.1007/978-3-031-64803-8_2

2.2 Software Reliability Modeling for Big Fault Data

Let $O(t)$ be the cumulative number of detected faults latent in the OSS by operation time t ($t \geq 0$) registered in the bug tracking system. Suppose that $O(t)$ takes on continuous real values. Since latent faults in the OSS are detected and eliminated during the operation phase, $O(t)$ gradually increases as the operational procedures go on. Thus, under common assumptions for *software reliability growth modeling* [1], the following linear differential equation can be formulated:

$$\frac{dO(t)}{dt} = b(t)\{a - O(t)\}, \tag{2.1}$$

where $b(t)$ is the software fault-detection rate at operation time t and a non-negative function, a is the number of latent faults in the OSS.

Considering the characteristic of the operation phase in OSS development environment, the software fault-reporting phenomena keep an irregular state in the operation phase, due to the network access and OSS users. Then, we consider the big data in order to assess the reliability for OSS computing. Therefore, we extend Eq. (2.1) to the following stochastic differential equation modeling considering a Brownian motion [13,14]:

$$\frac{dO(t)}{dt} = \{b(t) + \sigma v(t)\}\{a - O(t)\}, \tag{2.2}$$

where σ is a positive constant representing a magnitude of the irregular fluctuation, $v(t)$ a standardized Gaussian white noise. We assume that $O(t)$ is related with the software fault-detection rate $b(t)$ depending on the failure-occurrence phenomenon.

We extend Eq. (2.2) to the following stochastic differential equations of an Itô type [15]:

$$dO(t) = \left\{b(t) - \frac{1}{2}\sigma^2\right\}\{a - O(t)\}dt, \tag{2.3}$$

where $\omega(t)$ is the Wiener process which is formally defined as an integration of the white noise $v(t)$ with respect to time t. Then, a Wiener process, $\hat{\omega}(t)$, is a Gaussian process and has the following properties:

$$\Pr[\hat{\omega}(0) = 0] = 1, \tag{2.4}$$

$$\mathrm{E}[\hat{\omega}(t)] = 0, \tag{2.5}$$

$$\mathrm{E}[\hat{\omega}(t)\hat{\omega}(t')] = \mathrm{Min}[t, t'], \tag{2.6}$$

where $\Pr[\cdot]$ and $\mathrm{E}[\cdot]$ represent the probability and expectation, respectively.

By using Itô's formula [13,14], we can obtain the solution of Eq. (2.3) under the initial condition $O(0) = 0$ as follows [15]:

$$O(t) = a\left[1 - \exp\left\{-\int_0^t b(s)ds - \sigma\omega(t)\right\}\right]. \tag{2.7}$$

Using solution process $O(t)$ in Eq. (2.7), we can derive several software reliability measures.

2.4 Cost Optimization Based on the Stochastic Differential Equation Model

Moreover, we define the software fault-detection rate per fault in case of $b(t)$ defined as:

$$\int_0^t b(s)ds \doteq \frac{\frac{dI(t)}{dt}}{a - I(t)}$$
$$= \frac{1+c}{1+c \cdot \exp(-bt)}, \quad (2.8)$$

where $I(t)$ means the mean value functions for the inflection S-shaped SRGM, based on a nonhomogeneous Poisson process (NHPP) [1], a the expected number of latent faults for SRGM, and b the fault-detection rate per fault. Generally, the parameter c is defined as $\frac{(1-l)}{l}$. We define the parameter l as the mean value of network traffic density in this chapter.

Therefore, the solution process of cumulative number of detected faults is obtained as follows:

$$O(t) = \left[1 - \frac{1+c}{1+c \cdot \exp(-bt)} \cdot \exp\left\{-bt - \sigma\omega(t)\right\}\right]. \quad (2.9)$$

2.3 Reliability Assessment Measures

Information on the cumulative number of detected faults in the system is important to estimate the situation of the progress on the OSS operation procedures. Since $O(t)$ is a random variable in the proposed model, its expected value can be a useful measure. It can be derived from Eq. (2.7) as follows [15]:

$$E[O(t)] = a\left[1 - \exp\left\{-\int_0^t b(s)ds + \frac{\sigma^2}{2}t\right\}\right], \quad (2.10)$$

where $E[O(t)]$ is the expected number of faults detected up to time t. Also, the number of remaining faults and the expected number of remaining faults at time t can be obtained as follow, respectively:

$$N(t) = a \cdot \exp\left\{-\int_0^t b(s)ds - \sigma\omega(t)\right\}, \quad (2.11)$$

$$E[N(t)] = a \cdot \exp\left\{-\int_0^t b(s)ds + \frac{\sigma^2}{2}t\right\}. \quad (2.12)$$

2.4 Cost Optimization Based on the Stochastic Differential Equation Model

Several optimal software release problems considering software development process have been proposed by several researchers [16, 17]. Considering the characteristics of big data, it is interesting for the software developers to predict and estimate the time when we should stop operating in order to maintain the OSS system efficiently. We define the following cost parameters:

c_1: the operation cost per unit time,
c_2: the fixing cost per fault during the operation,
c_3: the maintenance cost per fault after the operation.

Then, the expected software cost in the operation of OSS system can be formulated as:

$$C_1(t) = c_1 t + c_2 \mathrm{E}[O(t)]. \tag{2.13}$$

Also, the expected software maintenance cost after the maintenance of OSS system is represented as follows:

$$C_2(t) = c_3 \{a - \mathrm{E}[O(t)]\}. \tag{2.14}$$

Consequently, from Eqs. (2.13) and (2.14), the total expected software maintenance cost is given by

$$C(t) = C_1(t) + C_2(t). \tag{2.15}$$

The optimum operation time t^* is obtained by minimizing $C(t)$ in Eq. (2.15). Moreover, we can derive the sample path of the total software cost based on Wiener process as follows:

$$C_w(t) = c_1 t + c_2 O(t) + c_3 \{a - O(t)\}. \tag{2.16}$$

2.5 Numerical Examples

The OSS is closely watched from the point of view of the cost reduction and the quick delivery. There are several open source projects in area of cloud computing. In particular, we focus on OpenStack [18] in order to evaluate the performance of our model. In this chapter, we show several numerical examples by using the data sets for OpenStack of cloud OSS. The data used in this chapter are collected in the bug tracking system on the website in terms of OpenStack OSS. Tables 2.1, 2.2, 2.3, 2.4, 2.5, 2.6, 2.7, 2.8, 2.9, 2.10, 2.11 and 2.12 show all logged on the bug tracking system of OpenStack.

2.5 Numerical Examples

Table 2.1 All data logged on the bug tracking system

Time	Fault	Time	Fault	Time	Fault	Time	Fault	Time	Fault	Time	Fault	Time	Fault
0	0	50	1	100	1	150	1	200	1	250	143	300	264
1	1	51	1	101	1	151	1	201	1	251	143	301	264
2	1	52	1	102	1	152	1	202	1	252	148	302	264
3	1	53	1	103	1	153	1	203	1	253	151	303	265
4	1	54	1	104	1	154	1	204	1	254	158	304	265
5	1	55	1	105	1	155	1	205	1	255	167	305	265
6	1	56	1	106	1	156	1	206	1	256	172	306	265
7	1	57	1	107	1	157	1	207	1	257	175	307	266
8	1	58	1	108	1	158	1	208	1	258	176	308	269
9	1	59	1	109	1	159	1	209	1	259	178	309	269
10	1	60	1	110	1	160	1	210	3	260	184	310	269
11	1	61	1	111	1	161	1	211	4	261	188	311	271
12	1	62	1	112	1	162	1	212	12	262	193	312	279
13	1	63	1	113	1	163	1	213	19	263	198	313	279
14	1	64	1	114	1	164	1	214	20	264	198	314	279
15	1	65	1	115	1	165	1	215	20	265	200	315	281
16	1	66	1	116	1	166	1	216	23	266	208	316	283
17	1	67	1	117	1	167	1	217	28	267	213	317	284
18	1	68	1	118	1	168	1	218	29	268	214	318	290
19	1	69	1	119	1	169	1	219	34	269	216	319	291
20	1	70	1	120	1	170	1	220	44	270	216	320	291
21	1	71	1	121	1	171	1	221	45	271	216	321	292
22	1	72	1	122	1	172	1	222	46	272	216	322	294
23	1	73	1	123	1	173	1	223	52	273	216	323	295
24	1	74	1	124	1	174	1	224	56	274	221	324	297
25	1	75	1	125	1	175	1	225	59	275	227	325	298
26	1	76	1	126	1	176	1	226	64	276	231	326	298
27	1	77	1	127	1	177	1	227	67	277	232	327	298
28	1	78	1	128	1	178	1	228	68	278	232	328	298
29	1	79	1	129	1	179	1	229	68	279	233	329	299
30	1	80	1	130	1	180	1	230	71	280	237	330	302
31	1	81	1	131	1	181	1	231	75	281	239	331	302
32	1	82	1	132	1	182	1	232	77	282	241	332	302
33	1	83	1	133	1	183	1	233	82	283	245	333	302
34	1	84	1	134	1	184	1	234	89	284	248	334	302
35	1	85	1	135	1	185	1	235	98	285	248	335	302
36	1	86	1	136	1	186	1	236	99	286	248	336	307
37	1	87	1	137	1	187	1	237	100	287	250	337	311
38	1	88	1	138	1	188	1	238	103	288	253	338	315
39	1	89	1	139	1	189	1	239	106	289	253	339	320
40	1	90	1	140	1	190	1	240	111	290	253	340	320
41	1	91	1	141	1	191	1	241	116	291	254	341	321
42	1	92	1	142	1	192	1	242	119	292	254	342	331
43	1	93	1	143	1	193	1	243	119	293	255	343	332
44	1	94	1	144	1	194	1	244	122	294	256	344	339
45	1	95	1	145	1	195	1	245	127	295	257	345	343
46	1	96	1	146	1	196	1	246	132	296	258	346	344
47	1	97	1	147	1	197	1	247	135	297	262	347	345
48	1	98	1	148	1	198	1	248	140	298	264	348	345
49	1	99	1	149	1	199	1	249	143	299	264	349	346

Table 2.2 All data logged on the bug tracking system (Cont. 1)

Time	Fault	Time	Fault	Time	Fault	Time	Fault	Time	Fault	Time	Fault
350	351	400	589	450	687	500	840	550	1365	600	1674	650	1858
351	355	401	590	451	691	501	842	551	1365	601	1675	651	1868
352	357	402	596	452	691	502	842	552	1367	602	1678	652	1877
353	365	403	597	453	691	503	842	553	1371	603	1682	653	1884
354	370	404	597	454	692	504	846	554	1377	604	1687	654	1888
355	371	405	597	455	697	505	847	555	1377	605	1689	655	1894
356	371	406	613	456	698	506	851	556	1400	606	1701	656	1894
357	381	407	615	457	699	507	853	557	1401	607	1706	657	1895
358	386	408	616	458	700	508	853	558	1402	608	1706	658	1899
359	389	409	617	459	700	509	855	559	1402	609	1711	659	1903
360	399	410	619	460	700	510	855	560	1403	610	1717	660	1923
361	399	411	619	461	701	511	858	561	1406	611	1728	661	1928
362	399	412	619	462	704	512	862	562	1406	612	1732	662	1933
363	400	413	621	463	707	513	875	563	1413	613	1734	663	1937
364	406	414	622	464	710	514	879	564	1418	614	1735	664	1937
365	407	415	622	465	716	515	892	565	1419	615	1737	665	1944
366	416	416	622	466	717	516	893	566	1420	616	1741	666	1946
367	431	417	622	467	717	517	894	567	1424	617	1749	667	1950
368	434	418	623	468	717	518	900	568	1437	618	1755	668	1957
369	434	419	623	469	718	519	907	569	1445	619	1757	669	1960
370	436	420	625	470	720	520	908	570	1453	620	1762	670	1961
371	440	421	628	471	722	521	916	571	1462	621	1762	671	1961
372	447	422	630	472	723	522	919	572	1465	622	1763	672	1963
373	492	423	633	473	724	523	919	573	1465	623	1769	673	1966
374	500	424	635	474	724	524	920	574	1467	624	1776	674	1970
375	500	425	635	475	725	525	925	575	1483	625	1781	675	1979
376	500	426	635	476	730	526	932	576	1489	626	1789	676	1982
377	500	427	638	477	741	527	933	577	1494	627	1793	677	1984
378	500	428	639	478	750	528	937	578	1498	628	1794	678	1988
379	500	429	643	479	754	529	938	579	1500	629	1794	679	1998
380	503	430	648	480	754	530	938	580	1502	630	1794	680	2000
381	511	431	649	481	754	531	938	581	1511	631	1797	681	2000
382	511	432	650	482	755	532	943	582	1543	632	1801	682	2004
383	511	433	650	483	755	533	948	583	1553	633	1807	683	2010
384	511	434	653	484	759	534	951	584	1561	634	1811	684	2014
385	511	435	658	485	762	535	964	585	1562	635	1811	685	2015
386	514	436	663	486	772	536	966	586	1564	636	1813	686	2016
387	515	437	665	487	776	537	966	587	1564	637	1821	687	2024
388	520	438	666	488	776	538	970	588	1573	638	1828	688	2026
389	521	439	666	489	781	539	977	589	1573	639	1832	689	2035
390	521	440	667	490	786	540	984	590	1578	640	1834	690	2046
391	528	441	669	491	792	541	991	591	1578	641	1835	691	2047
392	534	442	674	492	805	542	1304	592	1585	642	1835	692	2052
393	536	443	679	493	810	543	1333	593	1588	643	1835	693	2056
394	548	444	682	494	814	544	1333	594	1591	644	1838	694	2059
395	563	445	684	495	815	545	1336	595	1601	645	1849	695	2059
396	577	446	684	496	816	546	1339	596	1656	646	1853	696	2068
397	577	447	684	497	828	547	1348	597	1659	647	1856	697	2075
398	578	448	686	498	833	548	1354	598	1668	648	1857	698	2076
399	581	449	686	499	836	549	1364	599	1671	649	1858	699	2078

2.5 Numerical Examples

Table 2.3 All data logged on the bug tracking system (Cont. 2)

Time	Fault	Time	Fault	Time	Fault	Time	Fault	Time	Fault	Time	Fault	Time	Fault
700	2084	750	2430	800	2692	850	2894	900	3086	950	3264	1000	3523
701	2089	751	2463	801	2700	851	2898	901	3086	951	3267	1001	3530
702	2097	752	2479	802	2701	852	2900	902	3091	952	3269	1002	3531
703	2107	753	2497	803	2701	853	2902	903	3097	953	3275	1003	3550
704	2112	754	2500	804	2701	854	2910	904	3109	954	3283	1004	3561
705	2112	755	2500	805	2704	855	2916	905	3123	955	3287	1005	3561
706	2113	756	2504	806	2706	856	2924	906	3123	956	3289	1006	3561
707	2118	757	2511	807	2707	857	2927	907	3126	957	3290	1007	3563
708	2124	758	2520	808	2711	858	2930	908	3126	958	3292	1008	3565
709	2135	759	2526	809	2716	859	2933	909	3135	959	3307	1009	3568
710	2139	760	2531	810	2718	860	2934	910	3145	960	3312	1010	3580
711	2143	761	2534	811	2719	861	2943	911	3153	961	3316	1011	3583
712	2147	762	2538	812	2727	862	2948	912	3166	962	3370	1012	3584
713	2148	763	2543	813	2730	863	2948	913	3170	963	3372	1013	3584
714	2153	764	2550	814	2735	864	2953	914	3172	964	3372	1014	3585
715	2158	765	2555	815	2737	865	2954	915	3172	965	3373	1015	3586
716	2164	766	2560	816	2744	866	2956	916	3174	966	3378	1016	3589
717	2178	767	2564	817	2744	867	2957	917	3175	967	3385	1017	3593
718	2182	768	2564	818	2749	868	2959	918	3178	968	3392	1018	3595
719	2183	769	2564	819	2752	869	2963	919	3180	969	3404	1019	3596
720	2186	770	2567	820	2756	870	2963	920	3181	970	3413	1020	3596
721	2200	771	2569	821	2761	871	2964	921	3182	971	3413	1021	3597
722	2208	772	2574	822	2766	872	2965	922	3182	972	3414	1022	3597
723	2215	773	2581	823	2771	873	2965	923	3182	973	3420	1023	3598
724	2226	774	2584	824	2771	874	2965	924	3186	974	3424	1024	3599
725	2238	775	2584	825	2772	875	2968	925	3188	975	3426	1025	3610
726	2242	776	2584	826	2773	876	2977	926	3189	976	3431	1026	3615
727	2243	777	2587	827	2777	877	2989	927	3191	977	3433	1027	3615
728	2248	778	2591	828	2789	878	2993	928	3191	978	3433	1028	3615
729	2260	779	2611	829	2790	879	2997	929	3191	979	3437	1029	3615
730	2274	780	2617	830	2792	880	2999	930	3191	980	3442	1030	3622
731	2285	781	2623	831	2793	881	3000	931	3193	981	3442	1031	3626
732	2302	782	2623	832	2794	882	3002	932	3197	982	3449	1032	3635
733	2302	783	2623	833	2796	883	3005	933	3201	983	3452	1033	3639
734	2306	784	2626	834	2801	884	3015	934	3207	984	3454	1034	3639
735	2309	785	2636	835	2805	885	3018	935	3211	985	3454	1035	3640
736	2316	786	2639	836	2808	886	3020	936	3211	986	3458	1036	3644
737	2329	787	2645	837	2813	887	3024	937	3215	987	3460	1037	3647
738	2337	788	2651	838	2815	888	3028	938	3223	988	3464	1038	3649
739	2346	789	2653	839	2816	889	3033	939	3231	989	3465	1039	3652
740	2348	790	2654	840	2838	890	3039	940	3234	990	3478	1040	3654
741	2353	791	2655	841	2845	891	3048	941	3238	991	3488	1041	3654
742	2359	792	2659	842	2850	892	3050	942	3243	992	3489	1042	3655
743	2368	793	2663	843	2857	893	3051	943	3243	993	3490	1043	3659
744	2386	794	2667	844	2869	894	3051	944	3244	994	3495	1044	3667
745	2402	795	2672	845	2870	895	3053	945	3247	995	3510	1045	3673
746	2413	796	2677	846	2871	896	3054	946	3250	996	3514	1046	3675
747	2413	797	2677	847	2874	897	3056	947	3254	997	3519	1047	3676
748	2415	798	2678	848	2881	898	3068	948	3262	998	3521	1048	3676
749	2423	799	2686	849	2888	899	3080	949	3264	999	3522	1049	3677

Table 2.4 All data logged on the bug tracking system (Cont. 3)

Time	Fault	Time	Fault	Time	Fault	Time	Fault	Time	Fault	Time	Fault	Time	Fault
1050	3681	1100	3877	1150	4146	1200	4459	1250	4728	1300	5010	1350	5402
1051	3684	1101	3889	1151	4150	1201	4462	1251	4728	1301	5011	1351	5409
1052	3685	1102	3900	1152	4157	1202	4462	1252	4729	1302	5016	1352	5422
1053	3688	1103	3908	1153	4157	1203	4462	1253	4736	1303	5023	1353	5435
1054	3688	1104	3908	1154	4159	1204	4465	1254	4746	1304	5035	1354	5443
1055	3688	1105	3913	1155	4164	1205	4469	1255	4759	1305	5045	1355	5452
1056	3689	1106	3920	1156	4186	1206	4473	1256	4765	1306	5057	1356	5452
1057	3693	1107	3932	1157	4193	1207	4480	1257	4770	1307	5060	1357	5453
1058	3697	1108	3940	1158	4199	1208	4485	1258	4770	1308	5064	1358	5460
1059	3703	1109	3944	1159	4202	1209	4485	1259	4773	1309	5073	1359	5467
1060	3705	1110	3947	1160	4202	1210	4485	1260	4776	1310	5083	1360	5476
1061	3709	1111	3947	1161	4202	1211	4489	1261	4782	1311	5093	1361	5481
1062	3709	1112	3951	1162	4209	1212	4495	1262	4795	1312	5108	1362	5492
1063	3709	1113	3958	1163	4220	1213	4505	1263	4804	1313	5119	1363	5493
1064	3710	1114	3965	1164	4235	1214	4512	1264	4811	1314	5119	1364	5494
1065	3712	1115	3974	1165	4244	1215	4516	1265	4811	1315	5123	1365	5503
1066	3712	1116	3979	1166	4252	1216	4516	1266	4814	1316	5136	1366	5506
1067	3716	1117	3985	1167	4253	1217	4516	1267	4829	1317	5146	1367	5509
1068	3716	1118	3986	1168	4253	1218	4518	1268	4835	1318	5155	1368	5515
1069	3716	1119	3986	1169	4272	1219	4529	1269	4850	1319	5169	1369	5518
1070	3716	1120	3996	1170	4290	1220	4537	1270	4856	1320	5179	1370	5520
1071	3717	1121	4005	1171	4295	1221	4540	1271	4859	1321	5181	1371	5520
1072	3721	1122	4015	1172	4305	1222	4557	1272	4862	1322	5181	1372	5527
1073	3722	1123	4022	1173	4306	1223	4557	1273	4863	1323	5188	1373	5535
1074	3726	1124	4027	1174	4307	1224	4559	1274	4875	1324	5205	1374	5549
1075	3733	1125	4028	1175	4307	1225	4562	1275	4879	1325	5215	1375	5557
1076	3733	1126	4031	1176	4309	1226	4564	1276	4886	1326	5226	1376	5559
1077	3733	1127	4038	1177	4333	1227	4570	1277	4898	1327	5245	1377	5559
1078	3743	1128	4043	1178	4340	1228	4571	1278	4906	1328	5245	1378	5560
1079	3746	1129	4055	1179	4357	1229	4573	1279	4906	1329	5245	1379	5563
1080	3753	1130	4057	1180	4362	1230	4573	1280	4910	1330	5256	1380	5568
1081	3758	1131	4065	1181	4362	1231	4573	1281	4928	1331	5265	1381	5580
1082	3759	1132	4065	1182	4364	1232	4576	1282	4935	1332	5266	1382	5587
1083	3759	1133	4068	1183	4371	1233	4581	1283	4940	1333	5275	1383	5590
1084	3759	1134	4072	1184	4373	1234	4593	1284	4940	1334	5278	1384	5591
1085	3765	1135	4075	1185	4382	1235	4601	1285	4940	1335	5279	1385	5593
1086	3774	1136	4077	1186	4387	1236	4607	1286	4940	1336	5281	1386	5604
1087	3791	1137	4082	1187	4392	1237	4607	1287	4940	1337	5299	1387	5609
1088	3797	1138	4088	1188	4392	1238	4607	1288	4950	1338	5311	1388	5612
1089	3800	1139	4088	1189	4392	1239	4625	1289	4955	1339	5320	1389	5615
1090	3800	1140	4088	1190	4395	1240	4635	1290	4957	1340	5337	1390	5617
1091	3801	1141	4091	1191	4400	1241	4647	1291	4959	1341	5342	1391	5620
1092	3809	1142	4106	1192	4411	1242	4672	1292	4960	1342	5342	1392	5630
1093	3819	1143	4119	1193	4425	1243	4679	1293	4962	1343	5344	1393	5641
1094	3831	1144	4125	1194	4432	1244	4679	1294	4963	1344	5365	1394	5654
1095	3845	1145	4126	1195	4433	1245	4680	1295	4968	1345	5371	1395	5663
1096	3859	1146	4127	1196	4433	1246	4689	1296	4980	1346	5380	1396	5669
1097	3859	1147	4127	1197	4443	1247	4692	1297	4990	1347	5390	1397	5675
1098	3861	1148	4130	1198	4444	1248	4709	1298	5006	1348	5397	1398	5675
1099	3871	1149	4141	1199	4446	1249	4723	1299	5010	1349	5399	1399	5679

Table 2.5 All data logged on the bug tracking system (Cont. 4)

Time	Fault	Time	Fault	Time	Fault	Time	Fault	Time	Fault	Time	Fault
1400	5686	1450	5942	1500	6199	1550	6457	1600	6826	1650	7139	1700	7409
1401	5694	1451	5951	1501	6206	1551	6461	1601	6826	1651	7139	1701	7423
1402	5704	1452	5962	1502	6207	1552	6461	1602	6827	1652	7141	1702	7440
1403	5709	1453	5967	1503	6208	1553	6464	1603	6837	1653	7143	1703	7448
1404	5713	1454	5967	1504	6210	1554	6466	1604	6854	1654	7144	1704	7453
1405	5714	1455	5967	1505	6217	1555	6475	1605	6871	1655	7145	1705	7459
1406	5715	1456	5980	1506	6225	1556	6486	1606	6885	1656	7145	1706	7459
1407	5719	1457	5984	1507	6229	1557	6495	1607	6895	1657	7147	1707	7460
1408	5726	1458	5997	1508	6231	1558	6505	1608	6896	1658	7148	1708	7475
1409	5729	1459	6004	1509	6235	1559	6505	1609	6901	1659	7150	1709	7485
1410	5734	1460	6015	1510	6236	1560	6507	1610	6918	1660	7153	1710	7492
1411	5735	1461	6015	1511	6236	1561	6516	1611	6923	1661	7157	1711	7496
1412	5737	1462	6015	1512	6239	1562	6525	1612	6937	1662	7165	1712	7500
1413	5738	1463	6022	1513	6247	1563	6527	1613	6949	1663	7169	1713	7502
1414	5740	1464	6032	1514	6260	1564	6533	1614	6963	1664	7169	1714	7506
1415	5742	1465	6043	1515	6267	1565	6538	1615	6966	1665	7172	1715	7510
1416	5747	1466	6050	1516	6271	1566	6540	1616	6968	1666	7179	1716	7518
1417	5751	1467	6053	1517	6271	1567	6540	1617	6979	1667	7198	1717	7524
1418	5753	1468	6053	1518	6271	1568	6550	1618	6988	1668	7212	1718	7525
1419	5753	1469	6053	1519	6273	1569	6567	1619	6992	1669	7225	1719	7536
1420	5754	1470	6061	1520	6278	1570	6580	1620	7003	1670	7234	1720	7537
1421	5768	1471	6062	1521	6284	1571	6592	1621	7006	1671	7237	1721	7538
1422	5775	1472	6073	1522	6290	1572	6601	1622	7006	1672	7238	1722	7544
1423	5777	1473	6084	1523	6302	1573	6601	1623	7006	1673	7243	1723	7550
1424	5784	1474	6092	1524	6305	1574	6602	1624	7017	1674	7248	1724	7573
1425	5791	1475	6092	1525	6306	1575	6614	1625	7026	1675	7257	1725	7588
1426	5793	1476	6092	1526	6311	1576	6628	1626	7033	1676	7267	1726	7603
1427	5794	1477	6092	1527	6315	1577	6638	1627	7038	1677	7272	1727	7603
1428	5801	1478	6097	1528	6321	1578	6649	1628	7044	1678	7273	1728	7613
1429	5810	1479	6099	1529	6327	1579	6658	1629	7045	1679	7273	1729	7639
1430	5819	1480	6104	1530	6330	1580	6661	1630	7046	1680	7280	1730	7657
1431	5832	1481	6129	1531	6330	1581	6671	1631	7051	1681	7285	1731	7683
1432	5857	1482	6129	1532	6332	1582	6684	1632	7057	1682	7295	1732	7693
1433	5857	1483	6130	1533	6338	1583	6707	1633	7066	1683	7306	1733	7705
1434	5857	1484	6133	1534	6354	1584	6717	1634	7068	1684	7310	1734	7707
1435	5862	1485	6134	1535	6366	1585	6741	1635	7072	1685	7316	1735	7725
1436	5870	1486	6140	1536	6376	1586	6748	1636	7072	1686	7317	1736	7735
1437	5877	1487	6142	1537	6391	1587	6749	1637	7073	1687	7327	1737	7751
1438	5882	1488	6147	1538	6391	1588	6751	1638	7085	1688	7331	1738	7780
1439	5888	1489	6147	1539	6391	1589	6759	1639	7094	1689	7343	1739	7790
1440	5889	1490	6147	1540	6394	1590	6768	1640	7103	1690	7349	1740	7796
1441	5889	1491	6154	1541	6407	1591	6775	1641	7106	1691	7354	1741	7801
1442	5893	1492	6159	1542	6416	1592	6782	1642	7108	1692	7355	1742	7808
1443	5898	1493	6168	1543	6424	1593	6783	1643	7109	1693	7357	1743	7816
1444	5912	1494	6172	1544	6429	1594	6784	1644	7109	1694	7363	1744	7830
1445	5917	1495	6175	1545	6430	1595	6788	1645	7115	1695	7375	1745	7836
1446	5929	1496	6176	1546	6431	1596	6793	1646	7124	1696	7384	1746	7846
1447	5929	1497	6177	1547	6433	1597	6799	1647	7129	1697	7397	1747	7854
1448	5929	1498	6184	1548	6439	1598	6808	1648	7132	1698	7403	1748	7855
1449	5936	1499	6192	1549	6447	1599	6822	1649	7138	1699	7404	1749	7860

Table 2.6 All data logged on the bug tracking system (Cont. 5)

Time	Fault	Time	Fault	Time	Fault	Time	Fault	Time	Fault	Time	Fault	Time	Fault
1750	7871	1800	8196	1850	8611	1900	9081	1950	9509	2000	9906	2050	10247
1751	7879	1801	8208	1851	8620	1901	9093	1951	9509	2001	9906	2051	10257
1752	7885	1802	8215	1852	8623	1902	9095	1952	9509	2002	9912	2052	10262
1753	7896	1803	8223	1853	8624	1903	9099	1953	9530	2003	9918	2053	10270
1754	7904	1804	8224	1854	8626	1904	9113	1954	9544	2004	9925	2054	10285
1755	7904	1805	8226	1855	8644	1905	9133	1955	9555	2005	9939	2055	10294
1756	7906	1806	8236	1856	8661	1906	9161	1956	9570	2006	9949	2056	10295
1757	7924	1807	8246	1857	8668	1907	9174	1957	9573	2007	9951	2057	10302
1758	7933	1808	8263	1858	8673	1908	9183	1958	9573	2008	9954	2058	10307
1759	7945	1809	8272	1859	8683	1909	9184	1959	9574	2009	9966	2059	10321
1760	7954	1810	8282	1860	8683	1910	9187	1960	9583	2010	9977	2060	10343
1761	7954	1811	8282	1861	8684	1911	9192	1961	9596	2011	9986	2061	10354
1762	7954	1812	8283	1862	8689	1912	9207	1962	9601	2012	9999	2062	10366
1763	7955	1813	8294	1863	8704	1913	9213	1963	9615	2013	10001	2063	10368
1764	7956	1814	8299	1864	8716	1914	9225	1964	9624	2014	10003	2064	10370
1765	7966	1815	8307	1865	8722	1915	9236	1965	9624	2015	10005	2065	10381
1766	7979	1816	8319	1866	8735	1916	9237	1966	9629	2016	10006	2066	10395
1767	7987	1817	8323	1867	8737	1917	9241	1967	9642	2017	10007	2067	10415
1768	7992	1818	8325	1868	8738	1918	9247	1968	9650	2018	10009	2068	10439
1769	7993	1819	8325	1869	8750	1919	9270	1969	9667	2019	10009	2069	10454
1770	7995	1820	8327	1870	8762	1920	9281	1970	9680	2020	10009	2070	10454
1771	8006	1821	8346	1871	8776	1921	9291	1971	9700	2021	10009	2071	10456
1772	8018	1822	8355	1872	8800	1922	9299	1972	9700	2022	10011	2072	10469
1773	8029	1823	8362	1873	8823	1923	9301	1973	9703	2023	10011	2073	10486
1774	8046	1824	8369	1874	8823	1924	9307	1974	9715	2024	10019	2074	10515
1775	8051	1825	8370	1875	8825	1925	9317	1975	9724	2025	10023	2075	10530
1776	8052	1826	8373	1876	8835	1926	9328	1976	9735	2026	10031	2076	10536
1777	8052	1827	8382	1877	8845	1927	9336	1977	9749	2027	10037	2077	10537
1778	8054	1828	8391	1878	8853	1928	9346	1978	9755	2028	10048	2078	10538
1779	8057	1829	8405	1879	8869	1929	9357	1979	9756	2029	10049	2079	10549
1780	8064	1830	8417	1880	8877	1930	9357	1980	9759	2030	10053	2080	10576
1781	8074	1831	8425	1881	8879	1931	9360	1981	9762	2031	10065	2081	10583
1782	8079	1832	8426	1882	8879	1932	9367	1982	9774	2032	10078	2082	10591
1783	8080	1833	8426	1883	8906	1933	9376	1983	9784	2033	10090	2083	10600
1784	8080	1834	8439	1884	8926	1934	9380	1984	9797	2034	10104	2084	10600
1785	8081	1835	8472	1885	8935	1935	9389	1985	9805	2035	10105	2085	10600
1786	8083	1836	8483	1886	8948	1936	9394	1986	9806	2036	10105	2086	10608
1787	8087	1837	8495	1887	8955	1937	9395	1987	9812	2037	10119	2087	10618
1788	8094	1838	8504	1888	8955	1938	9395	1988	9821	2038	10128	2088	10634
1789	8099	1839	8505	1889	8956	1939	9397	1989	9830	2039	10154	2089	10658
1790	8099	1840	8507	1890	8970	1940	9408	1990	9840	2040	10166	2090	10671
1791	8103	1841	8513	1891	8979	1941	9414	1991	9849	2041	10178	2091	10678
1792	8110	1842	8518	1892	9002	1942	9428	1992	9860	2042	10179	2092	10681
1793	8120	1843	8537	1893	9019	1943	9435	1993	9861	2043	10179	2093	10690
1794	8123	1844	8543	1894	9024	1944	9437	1994	9863	2044	10194	2094	10709
1795	8134	1845	8551	1895	9024	1945	9439	1995	9871	2045	10201	2095	10720
1796	8168	1846	8552	1896	9025	1946	9455	1996	9877	2046	10215	2096	10741
1797	8168	1847	8557	1897	9038	1947	9463	1997	9890	2047	10231	2097	10761
1798	8169	1848	8569	1898	9051	1948	9477	1998	9899	2048	10244	2098	10761
1799	8177	1849	8587	1899	9062	1949	9495	1999	9905	2049	10246	2099	10762

2.5 Numerical Examples

Table 2.7 All data logged on the bug tracking system (Cont. 6)

Time	Fault	Time	Fault	Time	Fault	Time	Fault	Time	Fault	Time	Fault
2100	10779	2150	11289	2200	11799	2250	12216	2300	12716	2350	13216	2400	13584
2101	10794	2151	11303	2201	11804	2251	12230	2301	12720	2351	13218	2401	13587
2102	10811	2152	11311	2202	11812	2252	12230	2302	12724	2352	13244	2402	13594
2103	10827	2153	11320	2203	11814	2253	12234	2303	12733	2353	13262	2403	13603
2104	10840	2154	11321	2204	11814	2254	12245	2304	12751	2354	13277	2404	13616
2105	10840	2155	11326	2205	11821	2255	12255	2305	12766	2355	13291	2405	13625
2106	10842	2156	11339	2206	11830	2256	12271	2306	12792	2356	13301	2406	13625
2107	10856	2157	11352	2207	11840	2257	12277	2307	12802	2357	13304	2407	13626
2108	10870	2158	11365	2208	11850	2258	12280	2308	12803	2358	13307	2408	13634
2109	10892	2159	11381	2209	11858	2259	12280	2309	12808	2359	13321	2409	13650
2110	10908	2160	11384	2210	11858	2260	12283	2310	12830	2360	13338	2410	13667
2111	10914	2161	11385	2211	11863	2261	12302	2311	12845	2361	13355	2411	13680
2112	10917	2162	11386	2212	11875	2262	12327	2312	12862	2362	13364	2412	13705
2113	10920	2163	11391	2213	11884	2263	12353	2313	12888	2363	13376	2413	13706
2114	10927	2164	11413	2214	11890	2264	12383	2314	12896	2364	13376	2414	13708
2115	10935	2165	11420	2215	11902	2265	12410	2315	12896	2365	13378	2415	13716
2116	10946	2166	11429	2216	11912	2266	12411	2316	12896	2366	13384	2416	13724
2117	10954	2167	11438	2217	11914	2267	12412	2317	12912	2367	13403	2417	13741
2118	10960	2168	11441	2218	11917	2268	12422	2318	12927	2368	13424	2418	13755
2119	10962	2169	11447	2219	11927	2269	12446	2319	12939	2369	13438	2419	13759
2120	10965	2170	11453	2220	11939	2270	12457	2320	12953	2370	13449	2420	13759
2121	10977	2171	11460	2221	11954	2271	12470	2321	12964	2371	13449	2421	13760
2122	10988	2172	11474	2222	11970	2272	12478	2322	12964	2372	13449	2422	13778
2123	11005	2173	11482	2223	11975	2273	12480	2323	12971	2373	13459	2423	13784
2124	11020	2174	11485	2224	11975	2274	12480	2324	12985	2374	13462	2424	13795
2125	11032	2175	11487	2225	11979	2275	12487	2325	12991	2375	13477	2425	13807
2126	11034	2176	11490	2226	11987	2276	12499	2326	13003	2376	13486	2426	13817
2127	11041	2177	11500	2227	12001	2277	12506	2327	13011	2377	13496	2427	13818
2128	11055	2178	11513	2228	12028	2278	12512	2328	13022	2378	13497	2428	13820
2129	11070	2179	11519	2229	12043	2279	12522	2329	13022	2379	13503	2429	13826
2130	11091	2180	11530	2230	12046	2280	12523	2330	13023	2380	13503	2430	13840
2131	11096	2181	11535	2231	12047	2281	12526	2331	13037	2381	13503	2431	13853
2132	11110	2182	11535	2232	12050	2282	12537	2332	13053	2382	13505	2432	13859
2133	11112	2183	11536	2233	12064	2283	12545	2333	13074	2383	13511	2433	13866
2134	11117	2184	11555	2234	12074	2284	12553	2334	13094	2384	13512	2434	13869
2135	11136	2185	11571	2235	12087	2285	12565	2335	13103	2385	13512	2435	13870
2136	11152	2186	11580	2236	12099	2286	12581	2336	13103	2386	13515	2436	13883
2137	11169	2187	11602	2237	12122	2287	12581	2337	13104	2387	13516	2437	13898
2138	11185	2188	11606	2238	12124	2288	12582	2338	13116	2388	13516	2438	13909
2139	11197	2189	11606	2239	12124	2289	12592	2339	13125	2389	13518	2439	13923
2140	11202	2190	11608	2240	12131	2290	12598	2340	13138	2390	13524	2440	13923
2141	11206	2191	11620	2241	12146	2291	12610	2341	13146	2391	13528	2441	13925
2142	11218	2192	11629	2242	12160	2292	12615	2342	13153	2392	13529	2442	13927
2143	11226	2193	11637	2243	12172	2293	12634	2343	13153	2393	13531	2443	13939
2144	11244	2194	11754	2244	12181	2294	12634	2344	13155	2394	13541	2444	13946
2145	11255	2195	11768	2245	12181	2295	12636	2345	13162	2395	13546	2445	13953
2146	11264	2196	11769	2246	12182	2296	12644	2346	13175	2396	13556	2446	13961
2147	11264	2197	11769	2247	12190	2297	12657	2347	13197	2397	13568	2447	13969
2148	11265	2198	11781	2248	12201	2298	12679	2348	13212	2398	13581	2448	13972
2149	11278	2199	11789	2249	12207	2299	12697	2349	13215	2399	13583	2449	13977

Table 2.8 All data logged on the bug tracking system (Cont. 7)

Time	Fault	Time	Fault	Time	Fault	Time	Fault	Time	Fault	Time	Fault
2450	13983	2500	14419	2550	14963	2600	15348	2650	15653	2700	16031	2750	16353
2451	13993	2501	14423	2551	14981	2601	15358	2651	15654	2701	16032	2751	16355
2452	13996	2502	14438	2552	14989	2602	15358	2652	15655	2702	16042	2752	16355
2453	14004	2503	14450	2553	14989	2603	15359	2653	15668	2703	16048	2753	16355
2454	14010	2504	14451	2554	14991	2604	15370	2654	15676	2704	16058	2754	16359
2455	14010	2505	14453	2555	15003	2605	15381	2655	15683	2705	16069	2755	16362
2456	14010	2506	14464	2556	15015	2606	15387	2656	15691	2706	16075	2756	16362
2457	14023	2507	14472	2557	15040	2607	15395	2657	15706	2707	16077	2757	16364
2458	14031	2508	14479	2558	15045	2608	15399	2658	15710	2708	16078	2758	16368
2459	14043	2509	14491	2559	15050	2609	15399	2659	15710	2709	16085	2759	16371
2460	14050	2510	14499	2560	15054	2610	15399	2660	15718	2710	16092	2760	16380
2461	14058	2511	14500	2561	15054	2611	15411	2661	15724	2711	16098	2761	16385
2462	14058	2512	14500	2562	15062	2612	15422	2662	15740	2712	16109	2762	16391
2463	14065	2513	14515	2563	15072	2613	15429	2663	15764	2713	16116	2763	16392
2464	14083	2514	14524	2564	15080	2614	15437	2664	15776	2714	16117	2764	16392
2465	14094	2515	14535	2565	15094	2615	15445	2665	15776	2715	16118	2765	16403
2466	14106	2516	14545	2566	15103	2616	15445	2666	15777	2716	16122	2766	16418
2467	14127	2517	14551	2567	15103	2617	15446	2667	15788	2717	16130	2767	16425
2468	14151	2518	14553	2568	15105	2618	15450	2668	15794	2718	16149	2768	16439
2469	14151	2519	14555	2569	15109	2619	15455	2669	15806	2719	16161	2769	16447
2470	14152	2520	14569	2570	15121	2620	15463	2670	15820	2720	16163	2770	16448
2471	14164	2521	14589	2571	15136	2621	15469	2671	15832	2721	16164	2771	16449
2472	14180	2522	14624	2572	15141	2622	15481	2672	15832	2722	16168	2772	16461
2473	14190	2523	14639	2573	15151	2623	15483	2673	15833	2723	16176	2773	16469
2474	14204	2524	14649	2574	15151	2624	15483	2674	15845	2724	16199	2774	16488
2475	14213	2525	14653	2575	15154	2625	15489	2675	15857	2725	16207	2775	16499
2476	14214	2526	14656	2576	15167	2626	15499	2676	15869	2726	16236	2776	16502
2477	14217	2527	14685	2577	15182	2627	15508	2677	15874	2727	16242	2777	16502
2478	14230	2528	14702	2578	15194	2628	15518	2678	15884	2728	16242	2778	16503
2479	14241	2529	14712	2579	15204	2629	15527	2679	15884	2729	16245	2779	16515
2480	14254	2530	14731	2580	15215	2630	15527	2680	15885	2730	16246	2780	16538
2481	14270	2531	14751	2581	15215	2631	15529	2681	15895	2731	16252	2781	16549
2482	14289	2532	14751	2582	15217	2632	15533	2682	15902	2732	16267	2782	16553
2483	14289	2533	14756	2583	15225	2633	15538	2683	15911	2733	16276	2783	16558
2484	14289	2534	14765	2584	15234	2634	15546	2684	15918	2734	16288	2784	16558
2485	14302	2535	14789	2585	15242	2635	15564	2685	15927	2735	16288	2785	16559
2486	14316	2536	14803	2586	15264	2636	15570	2686	15929	2736	16288	2786	16568
2487	14335	2537	14815	2587	15269	2637	15571	2687	15930	2737	16302	2787	16578
2488	14345	2538	14823	2588	15269	2638	15571	2688	15936	2738	16314	2788	16586
2489	14352	2539	14825	2589	15271	2639	15579	2689	15941	2739	16326	2789	16592
2490	14352	2540	14828	2590	15278	2640	15586	2690	15957	2740	16335	2790	16598
2491	14353	2541	14840	2591	15282	2641	15589	2691	15966	2741	16338	2791	16600
2492	14363	2542	14871	2592	15289	2642	15600	2692	15975	2742	16339	2792	16601
2493	14378	2543	14883	2593	15299	2643	15605	2693	15977	2743	16340	2793	16609
2494	14384	2544	14890	2594	15307	2644	15606	2694	15977	2744	16344	2794	16620
2495	14405	2545	14901	2595	15307	2645	15606	2695	15989	2745	16348	2795	16628
2496	14407	2546	14903	2596	15308	2646	15618	2696	15998	2746	16351	2796	16635
2497	14407	2547	14903	2597	15315	2647	15628	2697	16017	2747	16353	2797	16652
2498	14407	2548	14920	2598	15320	2648	15639	2698	16027	2748	16353	2798	16652
2499	14415	2549	14928	2599	15333	2649	15650	2699	16031	2749	16353	2799	16663

2.5 Numerical Examples

Table 2.9 All data logged on the bug tracking system (Cont. 8)

Time	Fault	Time	Fault	Time	Fault	Time	Fault	Time	Fault	Time	Fault
2800	16676	2850	17011	2900	17398	2950	17718	3000	17979	3050	18211	3100	18469
2801	16684	2851	17020	2901	17403	2951	17721	3001	17981	3051	18212	3101	18477
2802	16702	2852	17032	2902	17411	2952	17721	3002	17981	3052	18223	3102	18482
2803	16715	2853	17032	2903	17411	2953	17728	3003	17984	3053	18229	3103	18488
2804	16731	2854	17032	2904	17412	2954	17735	3004	17985	3054	18239	3104	18493
2805	16733	2855	17035	2905	17417	2955	17740	3005	17993	3055	18250	3105	18496
2806	16734	2856	17042	2906	17422	2956	17748	3006	18005	3056	18254	3106	18496
2807	16738	2857	17055	2907	17434	2957	17750	3007	18009	3057	18254	3107	18496
2808	16744	2858	17074	2908	17443	2958	17757	3008	18009	3058	18255	3108	18499
2809	16754	2859	17119	2909	17447	2959	17758	3009	18011	3059	18265	3109	18505
2810	16765	2860	17125	2910	17449	2960	17759	3010	18015	3060	18273	3110	18508
2811	16779	2861	17140	2911	17449	2961	17773	3011	18020	3061	18277	3111	18512
2812	16780	2862	17143	2912	17454	2962	17777	3012	18024	3062	18287	3112	18513
2813	16782	2863	17149	2913	17470	2963	17780	3013	18033	3063	18288	3113	18514
2814	16787	2864	17158	2914	17487	2964	17789	3014	18036	3064	18289	3114	18514
2815	16794	2865	17168	2915	17499	2965	17795	3015	18036	3065	18292	3115	18515
2816	16803	2866	17181	2916	17502	2966	17795	3016	18036	3066	18302	3116	18516
2817	16815	2867	17191	2917	17502	2967	17795	3017	18041	3067	18308	3117	18520
2818	16818	2868	17191	2918	17502	2968	17800	3018	18048	3068	18315	3118	18521
2819	16819	2869	17191	2919	17510	2969	17803	3019	18060	3069	18322	3119	18521
2820	16821	2870	17199	2920	17519	2970	17807	3020	18065	3070	18330	3120	18521
2821	16829	2871	17213	2921	17531	2971	17816	3021	18070	3071	18331	3121	18522
2822	16839	2872	17223	2922	17551	2972	17823	3022	18071	3072	18333	3122	18523
2823	16849	2873	17231	2923	17564	2973	17823	3023	18072	3073	18334	3123	18527
2824	16857	2874	17233	2924	17565	2974	17823	3024	18083	3074	18339	3124	18531
2825	16863	2875	17233	2925	17566	2975	17832	3025	18088	3075	18354	3125	18534
2826	16872	2876	17235	2926	17575	2976	17843	3026	18097	3076	18363	3126	18536
2827	16874	2877	17245	2927	17580	2977	17848	3027	18103	3077	18369	3127	18536
2828	16883	2878	17258	2928	17591	2978	17857	3028	18112	3078	18370	3128	18537
2829	16892	2879	17268	2929	17600	2979	17859	3029	18112	3079	18372	3129	18542
2830	16895	2880	17279	2930	17609	2980	17860	3030	18112	3080	18377	3130	18545
2831	16906	2881	17284	2931	17610	2981	17860	3031	18120	3081	18385	3131	18546
2832	16921	2882	17284	2932	17611	2982	17866	3032	18127	3082	18392	3132	18555
2833	16921	2883	17286	2933	17626	2983	17875	3033	18137	3083	18395	3133	18563
2834	16922	2884	17291	2934	17640	2984	17888	3034	18139	3084	18402	3134	18563
2835	16926	2885	17301	2935	17651	2985	17897	3035	18145	3085	18403	3135	18563
2836	16933	2886	17315	2936	17665	2986	17903	3036	18145	3086	18403	3136	18572
2837	16939	2887	17320	2937	17668	2987	17904	3037	18147	3087	18406	3137	18576
2838	16944	2888	17323	2938	17668	2988	17904	3038	18155	3088	18412	3138	18584
2839	16951	2889	17324	2939	17669	2989	17912	3039	18159	3089	18416	3139	18591
2840	16953	2890	17326	2940	17674	2990	17916	3040	18171	3090	18424	3140	18594
2841	16953	2891	17332	2941	17680	2991	17925	3041	18177	3091	18433	3141	18594
2842	16965	2892	17341	2942	17688	2992	17932	3042	18180	3092	18433	3142	18595
2843	16970	2893	17346	2943	17691	2993	17940	3043	18180	3093	18435	3143	18600
2844	16980	2894	17353	2944	17696	2994	17943	3044	18183	3094	18446	3144	18607
2845	16990	2895	17362	2945	17696	2995	17944	3045	18186	3095	18450	3145	18612
2846	16993	2896	17363	2946	17698	2996	17946	3046	18194	3096	18456	3146	18617
2847	16996	2897	17367	2947	17704	2997	17958	3047	18198	3097	18462	3147	18619
2848	16996	2898	17374	2948	17710	2998	17967	3048	18202	3098	18468	3148	18619
2849	17005	2899	17390	2949	17714	2999	17974	3049	18207	3099	18468	3149	18620

Table 2.10 All data logged on the bug tracking system (Cont. 9)

Time	Fault	Time	Fault	Time	Fault	Time	Fault	Time	Fault	Time	Fault	Time	Fault
3150	18628	3200	18817	3250	19044	3300	19269	3350	19431	3400	19620	3450	19763
3151	18633	3201	18824	3251	19047	3301	19271	3351	19431	3401	19622	3451	19765
3152	18641	3202	18830	3252	19051	3302	19271	3352	19431	3402	19623	3452	19768
3153	18647	3203	18832	3253	19051	3303	19273	3353	19434	3403	19625	3453	19774
3154	18652	3204	18832	3254	19051	3304	19279	3354	19436	3404	19629	3454	19777
3155	18652	3205	18832	3255	19058	3305	19283	3355	19441	3405	19637	3455	19781
3156	18652	3206	18837	3256	19067	3306	19290	3356	19444	3406	19642	3456	19782
3157	18666	3207	18848	3257	19078	3307	19294	3357	19447	3407	19643	3457	19783
3158	18671	3208	18854	3258	19085	3308	19301	3358	19447	3408	19643	3458	19786
3159	18678	3209	18860	3259	19088	3309	19301	3359	19448	3409	19649	3459	19789
3160	18680	3210	18860	3260	19088	3310	19302	3360	19454	3410	19655	3460	19792
3161	18685	3211	18860	3261	19088	3311	19308	3361	19466	3411	19659	3461	19796
3162	18685	3212	18862	3262	19094	3312	19309	3362	19474	3412	19666	3462	19800
3163	18685	3213	18868	3263	19098	3313	19315	3363	19479	3413	19667	3463	19801
3164	18687	3214	18869	3264	19104	3314	19318	3364	19484	3414	19668	3464	19801
3165	18691	3215	18877	3265	19115	3315	19322	3365	19484	3415	19668	3465	19807
3166	18695	3216	18885	3266	19117	3316	19322	3366	19485	3416	19669	3466	19815
3167	18702	3217	18895	3267	19117	3317	19322	3367	19488	3417	19671	3467	19820
3168	18709	3218	18895	3268	19118	3318	19331	3368	19491	3418	19676	3468	19824
3169	18709	3219	18896	3269	19121	3319	19334	3369	19496	3419	19677	3469	19828
3170	18710	3220	18900	3270	19130	3320	19339	3370	19503	3420	19678	3470	19828
3171	18711	3221	18906	3271	19137	3321	19353	3371	19509	3421	19678	3471	19830
3172	18712	3222	18918	3272	19153	3322	19357	3372	19509	3422	19678	3472	19833
3173	18719	3223	18921	3273	19159	3323	19357	3373	19509	3423	19679	3473	19837
3174	18720	3224	18928	3274	19159	3324	19357	3374	19516	3424	19681	3474	19840
3175	18724	3225	18929	3275	19159	3325	19358	3375	19522	3425	19684	3475	19844
3176	18724	3226	18930	3276	19170	3326	19364	3376	19525	3426	19691	3476	19845
3177	18725	3227	18934	3277	19174	3327	19372	3377	19533	3427	19693	3477	19845
3178	18732	3228	18940	3278	19179	3328	19376	3378	19537	3428	19693	3478	19845
3179	18735	3229	18943	3279	19185	3329	19383	3379	19537	3429	19693	3479	19848
3180	18742	3230	18949	3280	19190	3330	19383	3380	19538	3430	19701	3480	19848
3181	18746	3231	18957	3281	19190	3331	19385	3381	19542	3431	19706	3481	19848
3182	18746	3232	18958	3282	19191	3332	19386	3382	19545	3432	19712	3482	19849
3183	18746	3233	18959	3283	19194	3333	19388	3383	19553	3433	19715	3483	19850
3184	18746	3234	18965	3284	19197	3334	19390	3384	19557	3434	19718	3484	19850
3185	18754	3235	18973	3285	19204	3335	19396	3385	19563	3435	19718	3485	19850
3186	18759	3236	18978	3286	19214	3336	19397	3386	19563	3436	19719	3486	19853
3187	18762	3237	18983	3287	19220	3337	19398	3387	19563	3437	19727	3487	19857
3188	18767	3238	18993	3288	19221	3338	19398	3388	19570	3438	19729	3488	19860
3189	18770	3239	18993	3289	19221	3339	19402	3389	19572	3439	19732	3489	19865
3190	18770	3240	18993	3290	19226	3340	19404	3390	19581	3440	19738	3490	19867
3191	18770	3241	19005	3291	19229	3341	19411	3391	19584	3441	19742	3491	19867
3192	18776	3242	19016	3292	19233	3342	19416	3392	19588	3442	19742	3492	19867
3193	18787	3243	19023	3293	19239	3343	19417	3393	19588	3443	19742	3493	19871
3194	18793	3244	19026	3294	19243	3344	19417	3394	19589	3444	19745	3494	19875
3195	18796	3245	19029	3295	19243	3345	19418	3395	19596	3445	19755	3495	19878
3196	18806	3246	19029	3296	19244	3346	19421	3396	19602	3446	19758	3496	19880
3197	18806	3247	19029	3297	19247	3347	19424	3397	19612	3447	19761	3497	19885
3198	18806	3248	19033	3298	19256	3348	19427	3398	19616	3448	19763	3498	19885
3199	18809	3249	19036	3299	19264	3349	19430	3399	19620	3449	19763	3499	19886

Table 2.11 All data logged on the bug tracking system (Cont. 10)

Time	Fault	Time	Fault	Time	Fault	Time	Fault	Time	Fault	Time	Fault
3500	19890	3550	20060	3600	20212	3650	20361	3700	20546	3750	20744	3800	20854
3501	19897	3551	20063	3601	20212	3651	20364	3701	20546	3751	20744	3801	20858
3502	19901	3552	20064	3602	20215	3652	20364	3702	20546	3752	20748	3802	20864
3503	19901	3553	20067	3603	20215	3653	20364	3703	20552	3753	20750	3803	20866
3504	19907	3554	20067	3604	20215	3654	20369	3704	20562	3754	20754	3804	20871
3505	19909	3555	20067	3605	20223	3655	20373	3705	20568	3755	20758	3805	20876
3506	19910	3556	20071	3606	20226	3656	20376	3706	20569	3756	20760	3806	20876
3507	19914	3557	20076	3607	20229	3657	20380	3707	20570	3757	20760	3807	20876
3508	19918	3558	20082	3608	20233	3658	20387	3708	20570	3758	20760	3808	20880
3509	19922	3559	20084	3609	20234	3659	20387	3709	20570	3759	20761	3809	20883
3510	19926	3560	20087	3610	20234	3660	20388	3710	20571	3760	20764	3810	20885
3511	19930	3561	20088	3611	20234	3661	20391	3711	20575	3761	20772	3811	20886
3512	19930	3562	20092	3612	20238	3662	20401	3712	20582	3762	20773	3812	20891
3513	19930	3563	20098	3613	20244	3663	20405	3713	20587	3763	20778	3813	20892
3514	19934	3564	20102	3614	20249	3664	20411	3714	20591	3764	20778	3814	20892
3515	19937	3565	20105	3615	20253	3665	20415	3715	20591	3765	20778	3815	20894
3516	19945	3566	20107	3616	20258	3666	20415	3716	20591	3766	20780	3816	20897
3517	19951	3567	20111	3617	20258	3667	20415	3717	20602	3767	20786	3817	20904
3518	19957	3568	20111	3618	20258	3668	20419	3718	20612	3768	20787	3818	20907
3519	19957	3569	20111	3619	20262	3669	20427	3719	20618	3769	20792	3819	20910
3520	19957	3570	20115	3620	20266	3670	20437	3720	20626	3770	20794	3820	20910
3521	19965	3571	20118	3621	20266	3671	20438	3721	20631	3771	20794	3821	20910
3522	19969	3572	20121	3622	20276	3672	20443	3722	20631	3772	20794	3822	20912
3523	19978	3573	20125	3623	20280	3673	20443	3723	20632	3773	20800	3823	20919
3524	19985	3574	20128	3624	20280	3674	20443	3724	20638	3774	20804	3824	20922
3525	19991	3575	20128	3625	20280	3675	20444	3725	20650	3775	20805	3825	20924
3526	19992	3576	20131	3626	20289	3676	20450	3726	20655	3776	20810	3826	20926
3527	19992	3577	20134	3627	20293	3677	20457	3727	20662	3777	20812	3827	20926
3528	19994	3578	20142	3628	20296	3678	20460	3728	20669	3778	20812	3828	20926
3529	20000	3579	20145	3629	20298	3679	20464	3729	20669	3779	20812	3829	20930
3530	20007	3580	20149	3630	20300	3680	20464	3730	20670	3780	20813	3830	20932
3531	20008	3581	20152	3631	20301	3681	20465	3731	20670	3781	20816	3831	20937
3532	20011	3582	20152	3632	20301	3682	20469	3732	20671	3782	20816	3832	20940
3533	20012	3583	20154	3633	20304	3683	20476	3733	20679	3783	20821	3833	20943
3534	20012	3584	20157	3634	20307	3684	20483	3734	20685	3784	20826	3834	20944
3535	20014	3585	20163	3635	20311	3685	20491	3735	20693	3785	20826	3835	20944
3536	20018	3586	20170	3636	20316	3686	20497	3736	20693	3786	20826	3836	20946
3537	20024	3587	20172	3637	20317	3687	20498	3737	20695	3787	20828	3837	20950
3538	20030	3588	20173	3638	20317	3688	20498	3738	20697	3788	20830	3838	20952
3539	20035	3589	20173	3639	20319	3689	20507	3739	20697	3789	20835	3839	20954
3540	20035	3590	20173	3640	20323	3690	20515	3740	20701	3790	20837	3840	20959
3541	20035	3591	20174	3641	20328	3691	20519	3741	20705	3791	20841	3841	20959
3542	20038	3592	20178	3642	20333	3692	20521	3742	20710	3792	20841	3842	20959
3543	20044	3593	20183	3643	20334	3693	20523	3743	20710	3793	20841	3843	20959
3544	20049	3594	20191	3644	20339	3694	20523	3744	20711	3794	20842	3844	20959
3545	20051	3595	20197	3645	20340	3695	20523	3745	20722	3795	20847	3845	20960
3546	20053	3596	20198	3646	20340	3696	20527	3746	20726	3796	20849	3846	20960
3547	20053	3597	20198	3647	20349	3697	20529	3747	20732	3797	20854	3847	20961
3548	20054	3598	20204	3648	20353	3698	20537	3748	20739	3798	20854	3848	20961
3549	20055	3599	20208	3649	20358	3699	20546	3749	20744	3799	20854	3849	20962

Table 2.12 All data logged on the bug tracking system (Cont. 11)

Time	Fault	Time	Fault	Time	Fault	Time	Fault	Time	Fault	Time	Fault
3850	20963	3900	21062	3950	21164	4000	21258	4050	21319	4100	21354
3851	20966	3901	21064	3951	21167	4001	21259	4051	21319	4101	21354
3852	20970	3902	21069	3952	21168	4002	21259	4052	21319	4102	21355
3853	20974	3903	21072	3953	21168	4003	21259	4053	21319		
3854	20975	3904	21072	3954	21168	4004	21260	4054	21322		
3855	20975	3905	21072	3955	21170	4005	21262	4055	21325		
3856	20975	3906	21077	3956	21173	4006	21262	4056	21325		
3857	20976	3907	21080	3957	21176	4007	21263	4057	21327		
3858	20978	3908	21082	3958	21180	4008	21265	4058	21329		
3859	20979	3909	21087	3959	21180	4009	21265	4059	21330		
3860	20980	3910	21090	3960	21180	4010	21266	4060	21331		
3861	20983	3911	21090	3961	21180	4011	21269	4061	21335		
3862	20983	3912	21091	3962	21181	4012	21270	4062	21336		
3863	20983	3913	21096	3963	21184	4013	21272	4063	21336		
3864	20984	3914	21102	3964	21184	4014	21273	4064	21336		
3865	20985	3915	21104	3965	21185	4015	21274	4065	21336		
3866	20988	3916	21107	3966	21186	4016	21274	4066	21336		
3867	20991	3917	21108	3967	21187	4017	21274	4067	21337		
3868	20993	3918	21108	3968	21187	4018	21279	4068	21338		
3869	20994	3919	21108	3969	21191	4019	21279	4069	21338		
3870	20996	3920	21110	3970	21195	4020	21280	4070	21338		
3871	20996	3921	21115	3971	21198	4021	21281	4071	21338		
3872	20998	3922	21121	3972	21199	4022	21281	4072	21338		
3873	21001	3923	21125	3973	21201	4023	21281	4073	21338		
3874	21004	3924	21128	3974	21202	4024	21281	4074	21338		
3875	21007	3925	21130	3975	21204	4025	21284	4075	21339		
3876	21007	3926	21130	3976	21206	4026	21285	4076	21339		
3877	21007	3927	21130	3977	21208	4027	21286	4077	21340		
3878	21010	3928	21132	3978	21210	4028	21291	4078	21341		
3879	21010	3929	21136	3979	21212	4029	21291	4079	21341		
3880	21016	3930	21139	3980	21214	4030	21291	4080	21341		
3881	21022	3931	21139	3981	21214	4031	21291	4081	21342		
3882	21028	3932	21139	3982	21214	4032	21291	4082	21343		
3883	21031	3933	21140	3983	21215	4033	21296	4083	21344		
3884	21032	3934	21144	3984	21217	4034	21301	4084	21346		
3885	21032	3935	21144	3985	21219	4035	21301	4085	21348		
3886	21034	3936	21148	3986	21223	4036	21303	4086	21348		
3887	21036	3937	21150	3987	21226	4037	21303	4087	21349		
3888	21041	3938	21151	3988	21226	4038	21303	4088	21350		
3889	21044	3939	21151	3989	21226	4039	21303	4089	21351		
3890	21044	3940	21151	3990	21231	4040	21306	4090	21353		
3891	21044	3941	21153	3991	21234	4041	21309	4091	21353		
3892	21045	3942	21155	3992	21236	4042	21310	4092	21353		
3893	21047	3943	21158	3993	21241	4043	21311	4093	21353		
3894	21052	3944	21158	3994	21242	4044	21311	4094	21353		
3895	21053	3945	21160	3995	21242	4045	21311	4095	21353		
3896	21056	3946	21160	3996	21243	4046	21312	4096	21353		
3897	21056	3947	21160	3997	21247	4047	21314	4097	21353		
3898	21056	3948	21161	3998	21250	4048	21315	4098	21353		
3899	21060	3949	21162	3999	21253	4049	21319	4099	21354		

2.6 Concluding Remarks

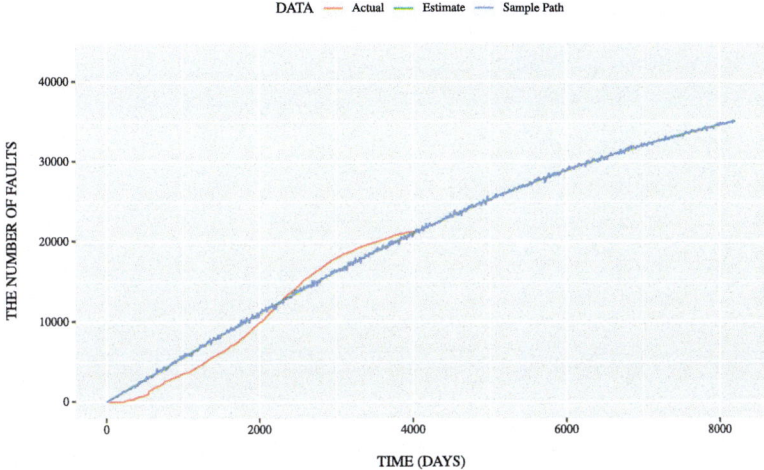

Fig. 2.1 The estimated expected number of detected faults

The estimated expected number of detected faults in Eq. (2.9) and the sample path of estimated number of detected faults in Eq. (2.10) are shown in Fig. 2.1. Also, the estimated expected number of remaining faults in Eq. (2.12) and the sample path of estimated number of remaining faults in Eq. (2.12) is shown in Fig. 2.2. The optimum maintenance time can be estimated as $t^* = 13874.1$ days (38.011 years) from Fig. 2.3.

Moreover, the estimated expected total software cost in Eq. (2.15) and the sample path of estimated total software cost in Eq. (2.16) are shown in Fig. 2.3.

From above mentioned results, we have found that our model can describe the characteristics of the OSS computing and fault big data. Our method will be useful to assess the dynamic reliability based on the fault big data on OSS computing.

2.6 Concluding Remarks

We have focused on the OSS for the big data as the next-generation software service paradigm. Also, we have discussed the method of reliability assessment considering the Wiener process.

The fluctuation of fault-detection per unit time becomes large, if the number of cumulative fault-detection grow. In this case, we can consider the OSS fault-detection phenomenon as the Wiener process.

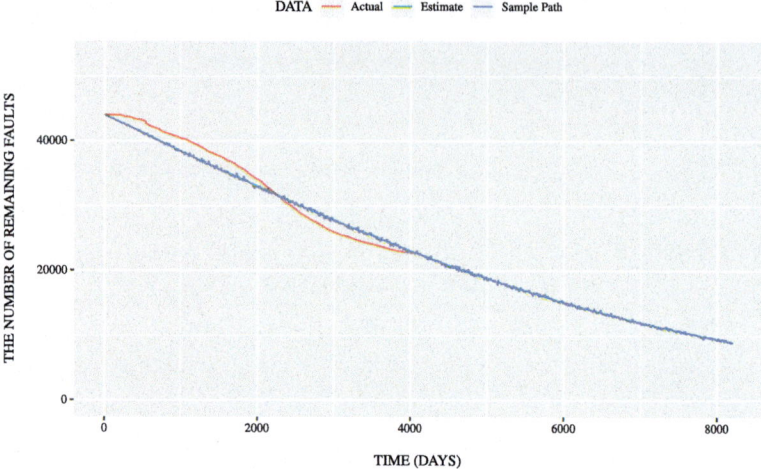

Fig. 2.2 The estimated expected number of remaining faults

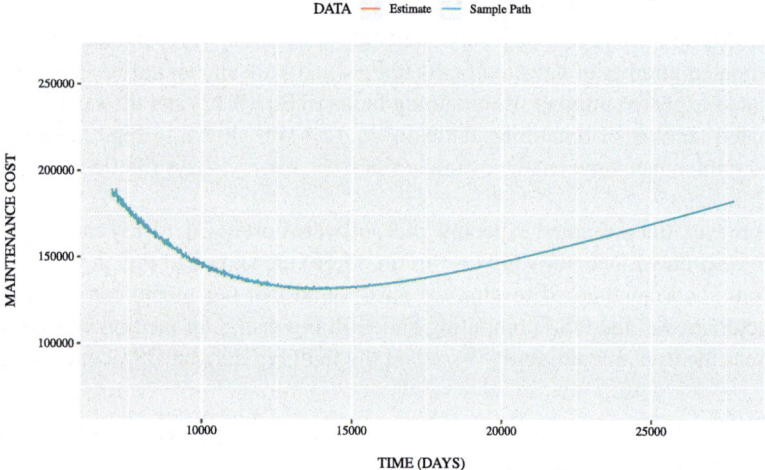

Fig. 2.3 The estimated total software cost

In this chapter, we have proposed OSS reliability analysis based on the stochastic differential equation model. Also, we have derived several reliability assessment measures from the proposed model. Moreover, we have analyzed actual data to show numerical examples of several software reliability assessment measures for the actual fault big data.

References

1. Yamada S (2014) Software reliability modeling: fundamentals and applications. Springer, Tokyo/Heidelberg
2. Lyu MR (ed) (1996) Handbook of software reliability engineering. IEEE Computer Society Press, Los Alamitos, CA, U.S.A
3. Musa JD, Iannino A, Okumoto K (1987) Software reliability: measurement, prediction, application. McGraw-Hill, New York
4. Kapur PK, Pham H, Gupta A, Jha PC (2011) Software reliability assessment with OR applications. Springer, London
5. Tamura Y, Yamada S (2021) Performance assessment based on stochastic differential equation and effort data for edge computing. J Softw Test Verification Reliab. Article first published online. https://doi.org/10.1002/stvr.1766. Wiley
6. Sone H, Tamura Y, Yamada S (2020) Stability assessment method considering fault fixing time in open source project. Int J Math Eng Manage Sci 5(4):591–601
7. Sone H, Tamura Y, Yamada S (2019) Statistical maintenance time estimation based on stochastic differential equation models in OSS development project. Comput Rev J PURKH 5:126–140
8. Sone H, Tamura Y, Yamada S (2019) A method of fault fix priority identification for open source project. Int J Recent Technol Eng 8(4):2396–2400. Blue Eyes Intelligence Engineering & Sciences Publication
9. Sone H, Tamura Y, Yamada S (2019) Prediction of fault fix time transition in large-scale open source project data. Data 4(3):1–12. Multidisciplinary Digital Publishing Institute, Switzerland. https://doi.org/10.3390/data4030109
10. Tamura Y, Sone H, Yamada S (2019) OSS project stability assessment support tool considering EVM based on Wiener process models. Appl Syst Innov 2(1):1–12. Multidisciplinary Digital Publishing Institute, Switzerland. https://doi.org/10.3390/asi2010001
11. Tamura Y, Yamada S (2018) Multi-dimensional software tool for OSS project management considering cloud with big data. Int J Reliab Qual Saf Eng 25(3):1850014-1–1850014-16. World Scientific
12. Tamura Y, Yamada S (2017) Open source software cost analysis with fault severity levels based on stochastic differential equation models. J Life Cycle Reliab Saf Eng 6(1):31–35. https://doi.org/10.1007/s41872-017-0009-5. Springer
13. Arnold L (1974) Stochastic differential equations-theory and applications. Wiley, New York
14. Wong E (1971) Stochastic processes in information and systems. McGraw-Hill, New York
15. Yamada S, Kimura M, Tanaka H, Osaki S (1994) Software reliability measurement and assessment with stochastic differential equations. IEICE Trans Fundam E77–A(1):109–116
16. Yamada S, Osaki S (1985) Cost-reliability optimal software release policies for software systems. IEEE Trans Reliab R-34(5):422–424
17. Yamada S, Osaki S (1987) Optimal software release policies with simultaneous cost and reliability requirements. Eur J Oper Res 31(1):46–51
18. The OpenStack project, Build the future of Open Infrastructure. https://www.openstack.org/

Two Dimensional Stochastic Differential Equation Model for OSS Reliability Analysis

3.1 Introduction

An OSS has been mainly used under the network service [1]. At present, the cloud computing service is changing to the edge computing [2–6]. The edge computing is useful for the continuity and responsibility of network service. Generally, it is well known that the fault-detection phenomenon depends on the maintenance effort, because the number of software faults is influenced by the effort expenditures. Our research group has proposed many effort maintenance models in the past [7–14]. All of these models have used the effort data sets of OSS. On the other hand, the OSS reliability assessment methods of this chapter are used the OSS fault count data. In the past, the software reliability has been generally assessed by using the SRGM's and the fault-detection count data. Then, the methods of software reliability assessment based on the SRGM's have been proposed by several researchers [15, 16].

Also, there are some interesting research works in terms of the cloud hardware, cloud service, mobile clouds, and cloud performance evaluation. However, most of them have focused on the case studies of cloud service and cloud data storage technologies. The effective methods of dynamic reliability assessment considering the environment such as cloud computing and OSS have been only a few presented in the past. In particular, it is very important to consider the status of fault-detection and big data in terms of the reliability assessment for cloud computing in the following standpoint:

- ▶ The data storage areas for cloud computing are reconfigured via the various mobile devices.
- ▶ The various mobile devices are connected via the network to the cloud service.
- ▶ The cloud computing has a particular maintenance phase such as the provisioning processes.

▶ The big data as the results from the huge and complex data by using the internet network cause the system-wide failures because of the complexity of data management.

From above situations, it is important to consider the indirect influences of big data on the reliability by using the effort data. We have proposed several methods of software reliability analysis for cloud computing in the past by using the effort data [7–14]. However, the effective methods of reliability assessment considering both the big data factor and fault one have been only a few presented, because it is very difficult to describe the indirect influence of big data and fault data as the reliability assessment measures.

Our research group has proposed the stochastic model for the effort control in the edge OSS operation in the past. We consider that the operation effort depends on various network environment of edge computing. In this chapter, we focus on the fault-detection count data from the actual cloud OSS. Then, we assume various network factors as two noises. In particular, we propose two dimensional stochastic differential equation models based on the cloud OSS fault-detection process in this chapter. Moreover, we show several numerical examples based on the proposed model.

3.2 Two Dimensional Wiener Process Modeling

Let $S(t)$ be the cumulative number of detected faults at the operation time t ($t \geq 0$) in the service of edge OSS. Suppose that $S(t)$ takes on continuous real values. Since the faults in the edge OSS is detected during the operation phase, $S(t)$ gradually increases as the operational procedures go on. Thus, under common assumptions for software reliability growth modeling [15], the following linear differential equation can be formulated:

$$\frac{dS(t)}{dt} = b(t)\{\alpha - S(t)\}, \quad (3.1)$$

where $b(t)$ is the OSS fault-detection rate at operation time t and a non-negative function, α, means the number of latent fault in the edge OSS.

Considering the characteristic of the operation phase in edge computing, the OSS fault-detection phenomena keep an irregular state in the operation phase, due to the network access of several edge servers. In case of the edge OSS, we have to consider that edge OSS depends on the number of edge servers, users, and the link speed of internet network such as 5G technology. In particular, the edge computing has the unique characteristics of edge servers. At present, the amount of data used by edge computing becomes large. Then, we consider the amount of detected faults depends on the reliability for edge computing. Therefore, we extend Eq. (3.1) to the following stochastic differential equation modeling considering two Brownian motions [13]:

3.2 Two Dimensional Wiener Process Modeling

$$\frac{dS_1(t)}{dt} = \{b_1(t) + \sigma_1 \rho_1(t)\}\{\alpha - S_1(t)\}, \tag{3.2}$$

$$\frac{dS_2(t)}{dt} = \{b_2(t) + \sigma_2 \rho_2(t)\}\{\alpha - S_2(t)\}, \tag{3.3}$$

where σ_1 and σ_2 are a positive constant representing a magnitude of the irregular fluctuation, respectively. Also, $\rho_1(t)$ and $\rho_2(t)$ a standardized Gaussian white noise, respectively. We assume that $S_1(t)$ is related with the OSS fault-detection rate $b_1(t)$ depending on the OSS fault-detection phenomenon. Similarly, $S_2(t)$ is related with the OSS fault-detection rate $b_2(t)$ depending on the network environment. We define the OSS fault-detection changing rate in case of $b(t)$ defined as:

$$\int_0^t b(s)ds \doteq \frac{\frac{dI(t)}{dt}}{\alpha - I(t)}$$
$$= \frac{1+c}{1 + c \cdot \exp(-bt)}, \tag{3.4}$$

where $I(t)$ means the mean value function for the inflection S-shaped SRGM, based on a nonhomogeneous Poisson process (NHPP) [15], α the number of latent fault, and b the fault-detection rate per fault. Generally, the parameter c is defined as the environment factor of edge computing in this paper.

Considering the independence of each noise, we can obtain the following integrated stochastic differential equation:

$$\frac{dS(t)}{dt} = \{b(t) + \sigma_1 \rho_1(t) + \sigma_2 \rho_2(t)\}\{\alpha - S(t)\}. \tag{3.5}$$

We extend the integrated two dimensional Eq. (3.5) to the following stochastic differential equation of an Itô type [13]:

$$dS_1(t) = \left\{b_1(t) - \frac{1}{2}\sigma_1^2\right\}\{\alpha - S_1(t)\}dt + \sigma_1\{\alpha - S_1(t)\}d\omega_1(t), \tag{3.6}$$

$$dS_2(t) = \left\{b_2(t) - \frac{1}{2}\sigma_2^2\right\}\{\alpha - S_2(t)\}dt + \sigma_2\{\alpha - S_2(t)\}d\omega_2(t), \tag{3.7}$$

where $\omega_i(t)$ is i-th one-dimensional Wiener process which is formally defined as an integration of the white noise $\rho_i(t)$ with respect to time t. Similarly, we can obtain the following integrated stochastic differential equation based on the independent of each noise.

$$dS(t) = \left\{b(t) - \frac{1}{2}(\sigma_1 + \sigma_2)^2\right\}\{\alpha - S(t)\}dt$$
$$+ \sigma_1\{\alpha - S(t)\}d\omega_1(t)$$
$$+ \sigma_2\{\alpha - S(t)\}d\omega_2(t). \tag{3.8}$$

We define the two dimensions processes $[\omega_1(t), \omega_2(t)]$ as follows [13]:

$$\widetilde{\omega}(t) = \left(\sigma_1^2 + \sigma_2^2\right)^{-\frac{1}{2}}\{\sigma_1 \omega_1(t) + \sigma_2 \omega_2(t)\}. \tag{3.9}$$

Then, two dimensional Wiener processes, $\widetilde{\omega}(t)$, are a Gaussian process and has the following properties:

$$\Pr[\widetilde{\omega}(0) = 0] = 1, \tag{3.10}$$

$$E[\widetilde{\omega}(t)] = 0, \tag{3.11}$$

$$E[\widetilde{\omega}(t)\widetilde{\omega}(t')] = \text{Min}[t, t'], \tag{3.12}$$

where $\Pr[\cdot]$ and $E[\cdot]$ represent the probability and expectation, respectively. Then, $\omega_1(t)$ and $\omega_2(t)$ are completely independent, respectively.

By using Itô's formula [17, 18], we can obtain the integrated solution of Eq. (3.8) under the initial condition $S(0) = 0$ as follows [19]:

$$S(t) = \alpha \left[1 - \exp\left\{ -\int_0^t b(s)ds - \sigma_1 \omega_1(t) - \sigma_2 \omega_2(t) \right\} \right]. \tag{3.13}$$

3.3 Cost Optimization Based on Two Dimensional Stochastic Differential Equation Model

Similarly, considering the characteristics of big fault data, it is interesting for the software developers to predict and estimate the time when we should stop operating in order to maintain a cloud software system efficiently. We define the following cost parameters [20, 21]:

c_1: the operation cost per unit time,
c_2: the fixing cost per fault during the operation,
c_3: the maintenance cost per fault after the operation.

Then, the expected software cost based on two noisy cases in the operation of cloud software system can be formulated as:

$$C_{s1}(t) = c_1 t + c_2 E[S(t)]. \tag{3.14}$$

Also, the expected software maintenance cost based on two noisy cases after the maintenance of cloud software system is represented as follows:

$$C_{s2}(t) = c_3 \{a - E[S(t)]\}. \tag{3.15}$$

Consequently, from Eqs. (3.14) and (3.15), the total expected software maintenance cost based on two noisy cases is given by

$$C_s(t) = C_{s1}(t) + C_{s2}(t). \tag{3.16}$$

The optimum maintenance time t^* is obtained by minimizing $C(t)$ in Eq. (3.16). Moreover, we can derive the sample path of the total software cost based on Wiener process as follows:

$$C_{ws}(t) = c_1 t + c_2 S(t) + c_3 \{a - S(t)\}. \tag{3.17}$$

3.4 Numerical Examples

We focus on *OpenStack* Project [22] including several edge components. In this chapter, we show numerical examples by using the data sets on the assumption of the edge OSS service. The data set used in this chapter is collected in the bug tracking system as shown in Tables 2.1, 2.2, 2.3, 2.4, 2.5, 2.6, 2.7, 2.8, 2.9, 2.10, 2.11, and 2.12.

The estimated sample path of the cumulative number of detected faults in Eq. (3.13) and the sample path of estimated number of detected faults are shown in Fig. 3.1. Also, the estimated sample path of the number of remaining faults is shown in Fig. 3.2.

Moreover, the sample path of the estimated total software cost in Eq. (3.17) are shown in Fig. 3.3. The optimum maintenance time can be estimated as $t^* = 13948.7$ days (38.216 years) from Fig. 3.3.

From above mentioned results, we have found that our model can describe the characteristics of edge computing according to the changes of the network environment by using two noises.

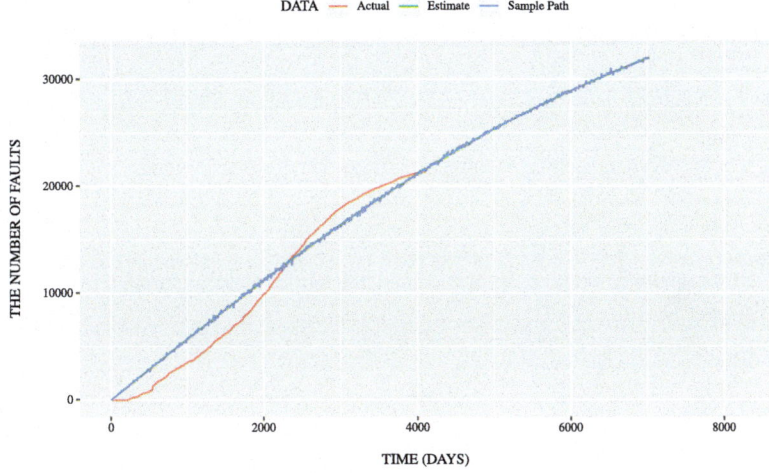

Fig. 3.1 The estimated sample path of the cumulative number of detected faults

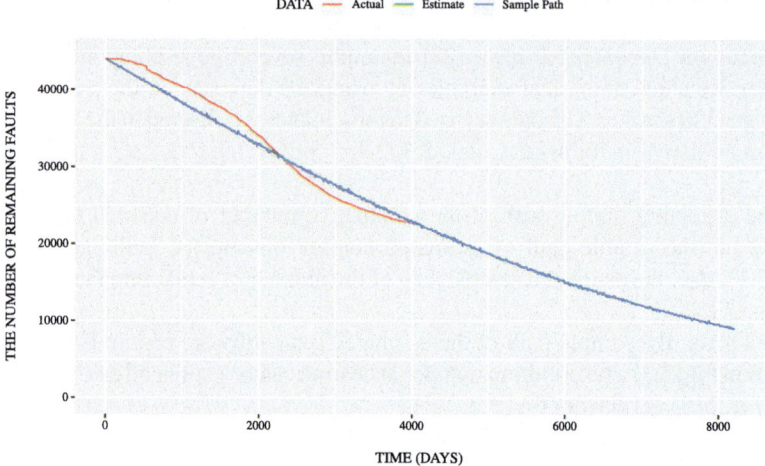

Fig. 3.2 The estimated sample path of the number of remaining faults

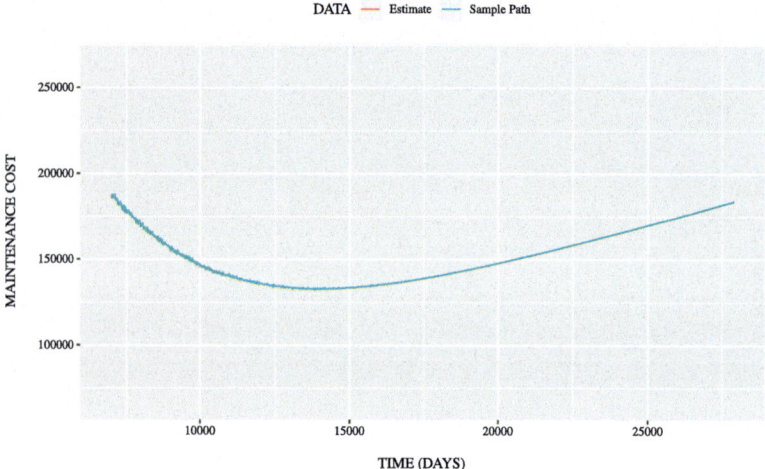

Fig. 3.3 The estimated total software cost

3.5 Concluding Remarks

The appropriate reliability control for the maintenance of edge software service will indirectly depends on the placement environment of edge server. Then, this chapter has proposed the two dimensional Wiener process model considering the irregular fluctuations as the characteristics of edge computing. It is difficult for the edge server managers to control the progress of server maintenance from the standpoint of the reliability. This chapter has discussed reliability assessment modeling by using the actual edge OSS fault data as follows:

◎ Edge OSS fault-detection phenomenon with time-variation
◎ Two dimensional Wiener processes

The proposed two dimensional noisy model will be helpful as the assessment measures of the reliability control for edge OSS service in operation phase.

References

1. Ibrahim IM et al (2018) A robust generic multi-authority attributes management system for cloud storage services. IEEE Trans Cloud Comput. https://doi.org/10.1109/TCC.2018.2867871
2. Ahmad AA et al (2019) Scalability analysis comparisons of cloud-based software services. J Cloud Comput Adv Syst Appl. https://doi.org/10.1186/s13677-019-0134-y
3. Ozcan MO, Odaci F, Ari I (2019) Remote debugging for containerized applications in edge computing environments. In: Proceedings of the 2019 IEEE international conference on edge computing (EDGE), Milan, Italy, pp 30–32. https://doi.org/10.1109/EDGE.2019.00021
4. Ngoko Y, Cérin C (2017) An edge computing platform for the detection of acoustic events. In: Proceedings of the 2017 IEEE international conference on edge computing (EDGE), Honolulu, HI, USA, pp 240–243, https://doi.org/10.1109/IEEE.EDGE.2017.44
5. Caprolu M, Di Pietro R, Lombardi F, Raponi S (2019) Edge computing perspectives: architectures, technologies, and open security issues. In: Proceedings of the 2019 IEEE international conference on edge computing (EDGE), Milan, Italy, pp 116–123. https://doi.org/10.1109/EDGE.2019.00035
6. Dolui K, Datta SK (2017) Comparison of edge computing implementations: fog computing, cloudlet and mobile edge computing. In: Proceedings of the 2017 global internet of things summit (GIoTS), Geneva, Switzerland, pp 1–6. https://doi.org/10.1109/GIOTS.2017.8016213
7. Tamura Y, Yamada S (2021) Performance assessment based on stochastic differential equation and effort data for edge computing. J Softw Test Verification Reliab, Article first published online: https://doi.org/10.1002/stvr.1766. Wiley
8. Sone H, Tamura Y, Yamada S (2020) Stability assessment method considering fault fixing time in open source project. Int J Math Eng Manage Sci 5(4):591–601
9. Sone H, Tamura Y, Yamada S (2019) Statistical maintenance time estimation based on stochastic differential equation models in OSS development project. Comput Rev J PURKH 5:126–140
10. Sone H, Tamura Y, Yamada S (2019) A method of fault fix priority identification for open source project. Int J Recent Technol Eng 8(4):2396–2400. Blue Eyes Intelligence Engineering & Sciences Publication
11. Sone H, Tamura Y, Yamada S (2019) Prediction of fault fix time transition in large-scale open source project data. Data 4(3):1–12. Multidisciplinary Digital Publishing Institute, Switzerland. https://doi.org/10.3390/data4030109
12. Tamura Y, Sone H, Yamada S (2019) OSS project stability assessment support tool considering EVM based on Wiener process models. Appl Syst Innov 2(1):1–12. Multidisciplinary Digital Publishing Institute, Switzerland. https://doi.org/10.3390/asi2010001
13. Tamura Y, Yamada S (2018) Multi-dimensional software tool for OSS project management considering cloud with big data. Int J Reliab Qual Saf Eng 25(3):1850014-1–1850014-16. World Scientific

14. Tamura Y, Yamada S (2017) Open source software cost analysis with fault severity levels based on stochastic differential equation models. J Life Cycle Reliab Saf Eng 6(1):31–35. https://doi.org/10.1007/s41872-017-0009-5. Springer
15. Yamada S (2014) Software reliability modeling: fundamentals and applications. Springer, Tokyo/Heidelberg
16. Kapur PK, Pham H, Gupta A, Jha PC (2011) Software reliability assessment with OR applications. Springer, London
17. Arnold L (1974) Stochastic differential equations-theory and applications. Wiley, New York
18. Wong E (1971) Stochastic processes in information and systems. McGraw-Hill, New York
19. Yamada S, Kimura M, Tanaka H, Osaki S (1994) Software reliability measurement and assessment with stochastic differential equations. IEICE Trans Fundam E77–A(1):109–116
20. Yamada S, Osaki S (1985) Cost-reliability optimal software release policies for software systems. IEEE Trans Reliab R-34(5):422–424
21. Yamada S, Osaki S (1987) Optimal software release policies with simultaneous cost and reliability requirements. Eur J Oper Res 31(1):46–51
22. The OpenStack project, Build the future of Open Infrastructure. https://www.openstack.org/

Jump Diffusion Process Model for OSS Reliability Analysis

4.1 Introduction

Many open source software (OSS) projects continues to increase around the world. Many OSS are operated under the irregular status of component-collision. In particular, several components of OSS will be made a change according to the progress of OSS development as time passes. Also, OSS include several versions such as bug-fix version, minor version, and major version, etc. Considering the characteristics of OSS projects, the operation performance of OSS development will take an irregular fluctuation during the operation, because several OSS components are updated by the OSS version-upgrade. In particular, the characteristics of OSS changes with the replacement of several components according to the numbers of several version upgrade. Several research works in terms of the reliability of OSS have been proposed by several researchers [1–3]. In the past, the methods of software reliability assessment based on the SRGM's have been proposed by several researchers [4,5].

This chapter proposes a method of OSS reliability assessment considering irregular fluctuation resulting from the characteristics with component dependency of OSS development and management. In particular, the OSS reliability assessment method based on the jump diffusion process model considering the component dependency in terms of OSS fault is proposed in order to comprehend the component characteristics of OSS in this chapter.

4.2 Jump Process Modeling

We apply a stochastic differential equation model to control the software reliability in the operational phase of OSS projects. Then, let $\Omega(t)$ be the cumulative number of detected faults up to operational time t ($t \geq 0$) in the operation of OSS project. Suppose that $\Omega(t)$ takes on continuous real values. Since the estimated amount of OSS faults are observed during the operational phase of the OSS project, $\Omega(t)$ gradually increases as the operational procedures go on. Based on software reliability growth modeling [4], the following linear differential equation in terms of OSS faults management can be formulated:

$$\frac{d\Omega(t)}{dt} = \beta(t)\{\alpha - \Omega(t)\}, \tag{4.1}$$

where $\beta(t)$ is the OSS fault-detection rate at OSS operational time t and a non-negative function, α, means the number of latent fault in the OSS operation.

Therefore, we extend Eq. (4.1) to the following stochastic differential equation with Brownian motion [6,7]:

$$\frac{d\Omega(t)}{dt} = \{\beta(t) + \sigma v(t)\}\{\alpha - \Omega(t)\}, \tag{4.2}$$

where σ is a positive constant representing a magnitude of the irregular fluctuation, and $v(t)$ a standardized Gaussian white noise. Then, we extend Eq. (4.2) to the following stochastic differential equation of an Itô type [6]:

$$d\Omega(t) = \left\{\beta(t) - \frac{1}{2}\sigma^2\right\}\{\alpha - \Omega(t)\}dt + \sigma\{\alpha - \Omega(t)\}d\omega(t), \tag{4.3}$$

where $\omega(t)$ is a one-dimensional Wiener process which is formally defined as an integration of the white noise $v(t)$ with respect to time t.

We can added the jump term to the proposed stochastic differential equation models in order to represent the irregular state around the time t influenced by various external factors in the development phase of OSS project. Then, the jump-diffusion process [8–13] is given as follows.

$$d\Omega_j(t) = \left\{\beta(t) - \frac{1}{2}\sigma^2\right\}\{\alpha - \Omega_j(t)\}dt \\ + \sigma\{\alpha - \Omega_j(t)\}d\omega(t) \\ + d\left\{\sum_{i=1}^{N_t(\rho)}(J_i - 1)\right\}, \tag{4.4}$$

where $N_t(\rho)$ is a Poisson point process with parameter ρ at operation time t. Also, $N_t(\rho)$ and ρ are defined as the number of occurred jumps and ρ the jump rate, respectively. $N_t(\rho)$, $\omega(t)$, and J_i are assumed to be mutually independent [8–13]. Moreover, J_i represents i-th jump range.

Generally, the jump diffusion models have been applied to the area of option pricing [14]. In particular, it is difficult to directly apply the idea of existing option pricing model to the development effort-expense phenomena, because the log-normal distribution is optimized for the option pricing area. Also, it is unnatural to apply the log-normal distribution based on the option pricing model to the OSS fault-detection phenomena, because it is usually assumed that the OSS fault-detection phenomena have non-biased distribution in the research area of software reliability engineering. Therefore, we define the following normal distribution function as Gaussian

Jump-diffusion process in order to consider the characteristics of OSS fault-detection phenomena:

$$J_i \equiv g_i(x) = \frac{1}{\sqrt{2\pi}\,\tau} \exp\left[-\frac{(x-\mu)^2}{2\tau^2}\right]. \tag{4.5}$$

where μ is the mean and τ is the standard deviation. Then, we assume that the i-th jump range J_i are independently estimated as the positive values in almost all cases, because the mean value μ keep a large value.

By using Itô's formula [6,7], the solution of the former equation can be obtained as follows:

$$\Omega_j(t) = \alpha\left[1 - \exp\left\{-\int_0^t b(s)ds - \sigma\omega(t) - \sum_{i=1}^{N_t(\rho)} \log J_i\right\}\right]. \tag{4.6}$$

4.3 Cost Optimization Based on Jump Diffusion Process Model

Similarly, considering the characteristics of big fault data and version-upgrade, it is interesting for the software developers to predict and estimate the time when we should stop operating in order to maintain the OSS operation efficiently. We define the following cost parameters:

c_1: the operation cost per unit time,
c_2: the fixing cost per fault during the operation,
c_3: the maintenance cost per fault after the operation.

Then, the software cost based on Wiener and jump cases in the operation of OSS operation can be formulated as:

$$C_{j1}(t) = c_1 t + c_2 \Omega_j(t). \tag{4.7}$$

Also, the software maintenance cost based on two noisy cases after the maintenance of cloud software system is represented as follows:

$$C_{j2}(t) = c_3\{\alpha - \Omega_j(t)\}. \tag{4.8}$$

Consequently, from Eqs. (4.7) and (4.8), the total software maintenance cost based on two noisy cases is given by

$$C_j(t) = C_{j1}(t) + C_{j2}(t). \tag{4.9}$$

The optimum maintenance time t^* is obtained by minimizing $C_j(t)$ in Eq. (4.9).

4.4 Numerical Examples

We focus on *OpenStack* Project [15] including several edge components. In this chapter, we show numerical examples by using the data sets on the assumption of the edge OSS service. The data set used in this chapter is collected in the bug tracking system as shown in Tables 2.1, 2.2, 2.3, 2.4, 2.5, 2.6, 2.7, 2.8, 2.9, 2.10, 2.11 and 2.12.

The estimated sample path of the cumulative number of detected faults is shown in Fig. 4.1. Also, the estimated sample path of the number of remaining faults is shown in Fig. 4.2. From Figs. 4.1 and 4.2, we find that the noise of jump difussion process becomes small according to the operating time procedures go on. On the other hand, the noise of Wiener process becomes no change according to the operating time procedures go on.

Moreover, the sample path of estimated total software cost in Eq. (4.9) are shown in Fig. 4.3. The optimum maintenance time can be estimated as $t^* = 13874.1$ days (38.011 years) from Fig. 3.3.

From above mentioned results, we have found that our jump diffusion process model can describe the characteristics of edge computing according to the changes of the network environment by using jump term.

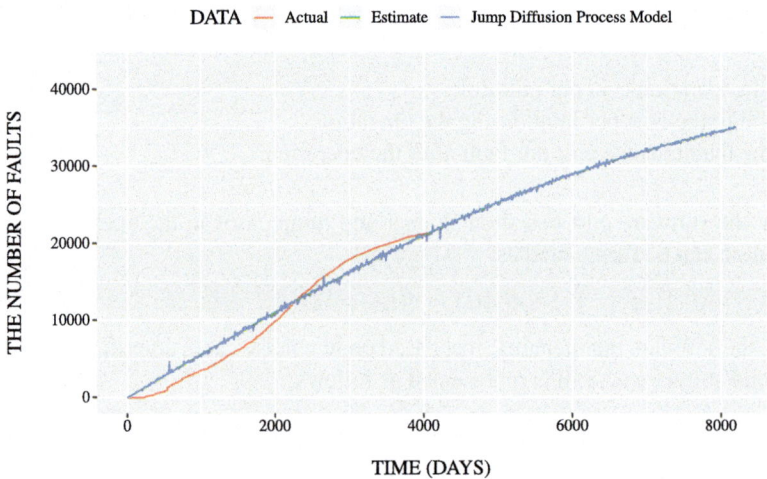

Fig. 4.1 The estimated expected number of detected faults

4.5 Concluding Remarks

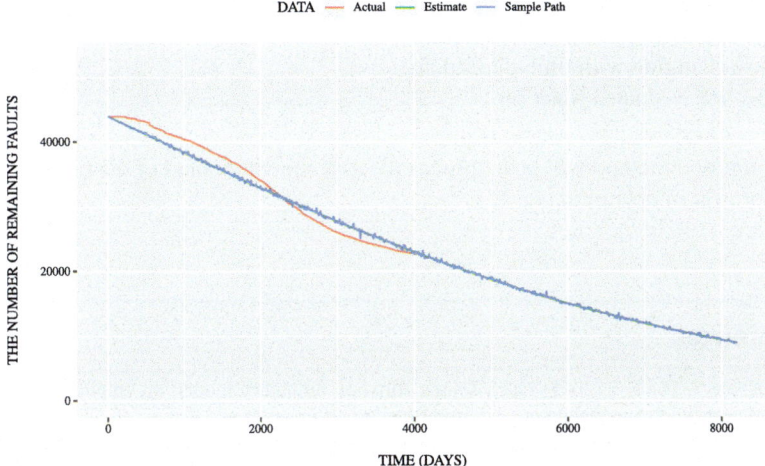

Fig. 4.2 The estimated expected number of remaining faults

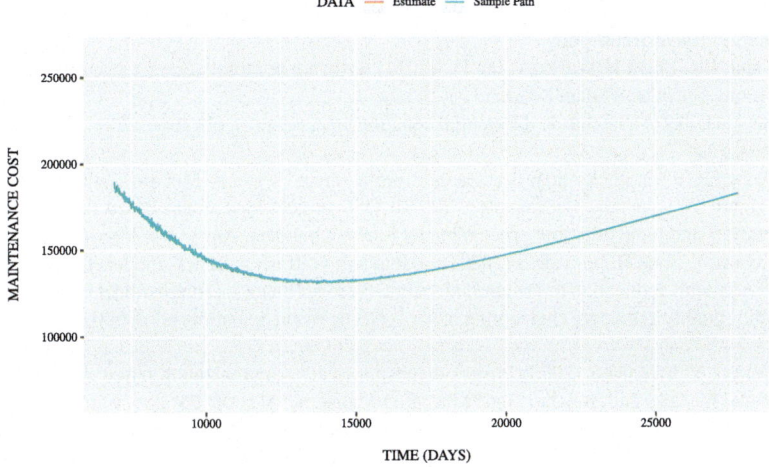

Fig. 4.3 The estimated total software cost

4.5 Concluding Remarks

The appropriate effort control for OSS development will indirectly depends on the usage quality, reliability, and cost reduction of OSS. This chapter has proposed a method of OSS project assessment considering the irregular fluctuations with jump and continuous noise from the non-contiguous characteristics of OSS development and management. It is difficult for the OSS project managers to manage the progress of OSS development considering the range of the non-contiguous prediction. This chapter has discussed by using the actual OSS fault data as follows:

- OSS component dependency with time-variation
- Wiener and jump diffusion processes
- Noise changing with time-dependent
- Inflection S-shaped function

The proposed method will be helpful as the assessment method of the reliability for OSS project in operation phase.

References

1. Zhoum Y, Davis J (2005) Open source software reliability model: an empirical approach. Proceed Workshop Open Source Softw Eng (WOSSE) 30(4):67–72
2. Li P, Shaw M, Herbsleb J, Ray B, Santhanam P (2004) Empirical evaluation of defect projection models for widely-deployed production software systems. In: Proceeding of the 12th international symposium on the foundations of software engineering (FSE–12). pp. 263–272
3. Norris J (2004) Mission-critical development with open source software. IEEE Softw Mag 21(1):42–49
4. Yamada S (2014) Software reliability modeling: fundamentals and applications. Springer-Verlag, Tokyo/Heidelberg
5. Kapur PK, Pham H, Gupta A, Jha PC (2011) Software reliability assessment with OR applications. Springer-Verlag, London
6. Arnold L (1974) Stochastic differential equations-theory and applications. John Wiley & Sons, New York
7. Wong E (1971) Stochastic processes in information and systems. McGraw-Hill, New York
8. Tamura Y, Watanabe H, Yamada S (2020) OSS project assessment based on discriminant analysis and jump diffusion process model for fault big data. Am J Oper Res 10(6):269–283
9. Tamura Y, Sone H, Yamada S (2020) Flexible jump diffusion process models for open source project with application to the optimal maintenance problem. In: International journal of reliability, quality and safety engineering, vol. 27, No. 6, World Scientific, pp. 2050020-1–2050020-18
10. Tamura Y, Yamada S (2020) Project maintenance effort optimization based on flexible JDP model for OSS fault big data. Int J Math Eng Manag Sci 5(1):66–75
11. Tamura Y, Sone H, Yamada S (2019) Productivity assessment based on jump diffusion model considering the effort management for OSS project. In: International journal of reliability, quality and safety engineering, vol. 26, No. 5, World Scientific, pp. 1950022-1–1950022-22
12. Tamura Y, Yamada S (2019) Maintenance effort management based on double jump diffusion model for OSS project. In: Annals of operations research, https://doi.org/10.1007/s10479-019-03170-w, Springer US, Online First, pp. 1–16
13. Tamura Y, Yamada S (2015) Reliability analysis based on a jump diffusion model with two Wiener processes for cloud computing with big data. In: Entropy, vol. 17, No. 7, Multidisciplinary Digital Publishing Institute, Switzerland, pp. 4533–4546
14. Merton RC (1976) Option pricing when underlying stock returns are discontinuous. J Finan Econ 3(1–2):125–144
15. The OpenStack project, Build the future of Open Infrastructure, https://www.openstack.org/

Cyclically Two Dimensional Stochastic Differential Equation Modeling

5.1 Introduction

At present, the cloud computing service has been mainly used as the network service [1]. The network service is changing from cloud computing to edge computing [2–6]. The edge computing is useful for the continuity and responsibility of network service. Also, OSS is embedded to the software system of edge computing. In the future, the edge computing will someday supersede cloud computing. Figure 5.1 shows the structure of edge computing and cloud one. Then, we define the wireless communication environment as white noises. Thereby, we can simultaneously consider the characteristics of edge and cloud, OSS, ubiquitous communication environment, and IoT.

However, the edge computing service requires a lot of effort for the operation. Generally, it is well known that the fault-detection phenomenon depends on the maintenance effort, because the number of detected software faults is influenced by the effort expenditure. Actually, several SRGM's with testing-effort have been proposed in the past [7]. We have focused on the effort expense management under the edge computing. Several open source software (OSS) are used in the situation of edge computing. Also, several research papers in terms of the reliability of OSS have been proposed by several researchers [8–10]. Then, the methods of software reliability assessment based on the SRGM's have been proposed by several researchers [11, 12]. On the other hand, it is important to appropriately control the operation effort in the case of the edge computing. The optimal control of operation effort for edge OSS computing will indirectly relate to the reliability and cost optimization.

In this chapter, we focus on the OSS development fault data set without effort data one in this chapter. This chapter proposes a stochastic model for the OSS reliability assessment in the edge OSS operation. We consider that the OSS reliability depends on various network environment of edge computing. Then, we assume various network factors as two kinds of noise. In particular, we define that these two noises have

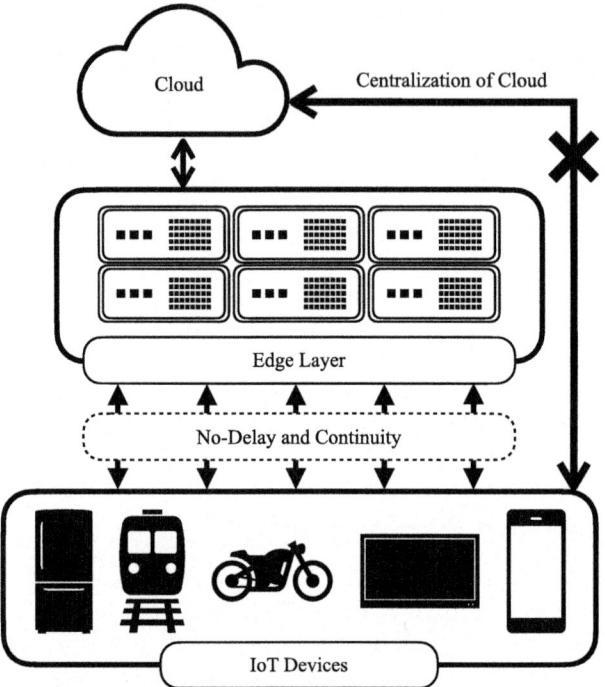

Fig. 5.1 Edge and cloud computing

the cyclic property from the standpoint of the network environment. Moreover, we show several numerical examples based on the proposed model.

5.2 OSS Reliability Assessment Modeling with Two Dimensional Cyclic Wiener Process

Let $F(t)$ be the cumulative number of detected faults at the operation time t ($t \geq 0$) in the service of edge OSS. Suppose that $F(t)$ takes on continuous real values. Since the number of OSS faults in the edge OSS is detected during the operation phase, $F(t)$ gradually increases as the operational procedures go on. Thus, under common assumptions for software reliability growth modeling [11], the following linear differential equation can be formulated:

$$\frac{dF(t)}{dt} = b(t)\{\alpha - F(t)\}, \tag{5.1}$$

where $b(t)$ is the OSS fault-detection rate at operation time t and a non-negative function, α, means the number of latent fault in the OSS operation.

Considering the characteristic of the operation phase in edge computing, the OSS fault-detection phenomena keep an irregular state in the operation phase, due to the

5.2 OSS Reliability Assessment Modeling with Two Dimensional Cyclic Wiener Process

network access of several edge servers. In case of edge OSS, we have to consider that edge OSS depends on the number of edge servers, users, and the link speed of internet network such as 5G technology. In particular, the edge computing has the unique characteristics of edge servers. At present, the amount of data used by edge computing becomes large. Then, we consider the fault-detection data in order to assess the reliability for edge computing. Therefore, we extend Eq. (5.1) to the following stochastic differential equation modeling considering two Brownian motions [13]:

$$\frac{dF_1(t)}{dt} = \{b_1(t) + \sigma_1 \rho_1(t)\}\{\alpha - F_1(t)\}, \quad (5.2)$$

$$\frac{dF_2(t)}{dt} = \{b_2(t) + \sigma_2 \rho_2(t)\}\{\alpha - F_2(t)\}, \quad (5.3)$$

where σ_1 and σ_2 are a positive constant representing a magnitude of the irregular fluctuation, $\rho_1(t)$ and $\rho_2(t)$ a standardized Gaussian white noise. We assume that $F_1(t)$ is related with the OSS fault-detection rate $b_1(t)$ depending on the OSS fault-detection phenomenon. Similarly, $F_2(t)$ is related with the OSS fault-detection rate $b_2(t)$ depending on the network environment. We define the OSS fault-detection rate in case of $b(t)$ defined as

$$\int_0^t b(s)ds \doteq \frac{\frac{dI(t)}{dt}}{\alpha - I(t)}$$
$$= \frac{1+c}{1+c \cdot \exp(-bt)}, \quad (5.4)$$

where $I(t)$ means the mean value function for the inflection S-shaped SRGM, based on a nonhomogeneous Poisson process (NHPP) [11], α the number of latent faults for OSS operation, and b the OSS fault-detection rate per fault. Generally, the parameter c is defined as the environment factor of edge computing in this chapter.

Considering the independence of each noise, we can obtain the following integrated stochastic differential equation:

$$\frac{dF(t)}{dt} = \{b(t) + \sigma_1 \rho_1(t) + \sigma_2 \rho_2(t)\}\{\alpha - F(t)\}. \quad (5.5)$$

We extend the integrated two dimensional Eq. (5.5) to the following stochastic differential equation of an Itô type [14]:

$$dF_1(t) = \left\{b_1(t) - \frac{1}{2}\sigma_1^2\right\}\{\alpha - F_1(t)\}dt + \sigma_1\{\alpha - F_1(t)\}d\omega_1(t), \quad (5.6)$$

$$dF_2(t) = \left\{b_2(t) - \frac{1}{2}\sigma_2^2\right\}\{\alpha - F_2(t)\}dt + \sigma_2\{\alpha - F_2(t)\}d\omega_2(t), \quad (5.7)$$

where $\omega_i(t)$ is i-th one-dimensional Wiener process which is formally defined as an integration of the white noise $\rho_i(t)$ with respect to time t. Similarly, we can obtain the following integrated stochastic differential equation based on the independent of each noise.

$$dF(t) = \left\{ b(t) - \frac{1}{2}(\sigma_1 + \sigma_2)^2 \right\} \{\alpha - F(t)\}dt$$
$$+ \sigma_1\{\alpha - F(t)\}d\omega_1(t)$$
$$+ \sigma_2\{\alpha - F(t)\}d\omega_2(t). \quad (5.8)$$

We define the two dimensions processes $[\omega_1(t), \omega_2(t)]$ as follows [13]:

$$\widetilde{\omega}(t) = \left(\sigma_1^2 + \sigma_2^2\right)^{-\frac{1}{2}} \{\sigma_1\omega_1(t) + \sigma_2\omega_2(t)\}. \quad (5.9)$$

Then, two dimensional Wiener processes, $\widetilde{\omega}(t)$, are a Gaussian process and has the following properties:

$$\Pr[\widetilde{\omega}(0) = 0] = 1, \quad (5.10)$$

$$E[\widetilde{\omega}(t)] = 0, \quad (5.11)$$

$$E[\widetilde{\omega}(t)\widetilde{\omega}(t')] = \text{Min}[t, t'], \quad (5.12)$$

where $\Pr[\cdot]$ and $E[\cdot]$ represent the probability and expectation, respectively. Then, $\omega_1(t)$ and $\omega_2(t)$ are completely independent, respectively.

By using Itô's formula [15, 16], we can obtain the integrated solution of Eq. (5.8) under the initial condition $F(0) = 0$ as follows [14]:

$$F(t) = \alpha\left[1 - \exp\left\{-\int_0^t b(s)ds - \sigma_1\omega_1(t) - \sigma_2\omega_2(t)\right\}\right]. \quad (5.13)$$

Therefore, the cumulative number of detected faults of edge operation are obtained as follows:

$$F(t) = \alpha\left[1 - \frac{1+c}{1+c\cdot\exp(-bt)}\right.$$
$$\left.\cdot\exp\left\{-bt - \sigma_1\omega_1(t) - \sigma_2\omega_2(t)\right\}\right]. \quad (5.14)$$

Moreover, we consider the cyclic noises depended on the network service of edge computing. In the cases of $dF_{c1}(t) = dF_1(t)$ and $dF_{c2}(t) = dF_2(t)$, we define as follows:

$$dF_{c1}(t) = \left\{b_1(t) - \frac{1}{2}\sigma_1^2\right\}\{\alpha - F_{c1}(t)\}dt$$
$$+ p_*(t)\sigma_1\{\alpha - F_{c1}(t)\}d\omega_1(t), \quad (5.15)$$

$$dF_{c2}(t) = \left\{b_2(t) - \frac{1}{2}\sigma_2^2\right\}\{\alpha - F_{c2}(t)\}dt$$
$$+ q_*(t)\sigma_2\{\alpha - F_{c2}(t)\}d\omega_2(t), \quad (5.16)$$

5.2 OSS Reliability Assessment Modeling with Two Dimensional Cyclic Wiener Process

In the proposed model, we define the weight functions based on the Fourier series expansion as follows:

$$p_*(t) = \frac{a_0}{2} + \sum_{n=1}^{\infty}\left[a_n \cos\left(\frac{2\pi nt}{T}\right) + b_n \sin\left(\frac{2\pi nt}{T}\right)\right], \quad (5.17)$$

$$q_*(t) = \frac{a_0}{2} - \sum_{n=1}^{\infty}\left[a_n \cos\left(\frac{2\pi nt}{T}\right) - b_n \sin\left(\frac{2\pi nt}{T}\right)\right], \quad (5.18)$$

where n is the cyclic number, T is the number of actual data. In this chapter, we consider two cases of sawtooth and square waves. In the case of sawtooth wave, we can derive as follows:

$$a_0 = 0, \quad a_n = 0, \quad b_n = \frac{2}{n\pi}.$$

Therefore, $p_1(t)$ and $q_1(t)$ in case of sawtooth wave are define as follows:

$$p_1(t) = \sum_{n=1}^{\infty}\frac{2}{n\pi}\sin\left(\frac{2\pi nt}{T}\right), \quad (5.19)$$

$$q_1(t) = -\sum_{n=1}^{\infty}\frac{2}{n\pi}\sin\left(\frac{2\pi nt}{T}\right). \quad (5.20)$$

In the case of square wave, we can derive as follows:

$$a_0 = 0, \quad a_n = 0, \quad b_n = \frac{2}{n\pi}\{1 - (-1)^n\}.$$

Therefore, $p_2(t)$ and $q_2(t)$ in case of square wave are define as follows:

$$p_2(t) = \sum_{n=1}^{\infty}\frac{2}{n\pi}\{1 - (-1)^n\}\sin\left(\frac{2\pi nt}{T}\right), \quad (5.21)$$

$$q_2(t) = -\sum_{n=1}^{\infty}\frac{2}{n\pi}\{1 - (-1)^n\}\sin\left(\frac{2\pi nt}{T}\right). \quad (5.22)$$

We assume several conditions of edge network environment. Then, Eqs. (5.17) and (5.18) assume the vertically symmetrical functions for Eqs. (5.19) and (5.20), and Eqs. (5.21) and (5.22), respectively. Considering the research area of communication engineering, the mathematical technique based on Fourier series expansion is frequently used from the standpoint of network engineering. In this chapter, we introduce the concept of Fourier series expansion to the proposed noisy models. Thereby, we can propose the noisy models from the actual situation considering the network engineering.

Then, we can consider the following equation according to the independency of random variable.

$$\begin{aligned}d\{F_{c1}(t) + F_{c2}(t)\} =& \left\{b(t) - \frac{1}{2}(\sigma_1 + \sigma_2)^2\right\}\{\alpha_1 + \alpha_2 - C_1(t) - F_{c2}(t)\}dt \\&+ p_*(t)\sigma_1\{\alpha_1 + \alpha_2 - F_{c1}(t) - F_{c2}(t)\}d\omega_1(t) \\&+ q_*(t)\sigma_2\{\alpha_1 + \alpha_2 - F_{c1}(t) - F_{c2}(t)\}d\omega_2(t).\end{aligned} \quad (5.23)$$

Therefore, we can obtain the following integrated stochastic differential equation based on the independent of each noise under $\alpha = \alpha_1 + \alpha_2$ and $F_c(t) = F_{c1}(t) + F_{c2}(t)$.

$$dF_c(t) = \left\{ b(t) - \frac{1}{2}(\sigma_1 + \sigma_2)^2 \right\} \{\alpha - F_c(t)\}dt$$
$$+ p_*(t)\sigma_1\{\alpha - C(t)\}d\omega_1(t) + q_*(t)\sigma_2\{\alpha - F_c(t)\}d\omega_2(t). \quad (5.24)$$

We define the two dimensions processes $[\omega_1(t), \omega_2(t)]$ as follows [15, 16]:

$$\widetilde{\omega}(t) = \left(\sigma_1^2 + \sigma_2^2\right)^{-\frac{1}{2}} \{\sigma_1\omega_1(t) + \sigma_2\omega_2(t)\}. \quad (5.25)$$

Then, two dimensional Wiener processes, $\widetilde{\omega}(t)$, are a Gaussian process and has the following properties:

$$\Pr[\widetilde{\omega}(0) = 0] = 1, \quad (5.26)$$

$$E[\widetilde{\omega}(t)] = 0, \quad (5.27)$$

$$E[\widetilde{\omega}(t)\widetilde{\omega}(t')] = \text{Min}[t, t'], \quad (5.28)$$

where $\Pr[\cdot]$ and $E[\cdot]$ represent the probability and expectation, respectively. Then, $\omega_1(t)$ and $\omega_2(t)$ are completely independent, respectively.

Moreover, we define the software fault-detection rate per fault in case of $b(t)$ defined as:

$$\int_0^t b(s)ds \doteq \frac{\frac{dI(t)}{dt}}{a - I(t)}$$
$$= \frac{1+c}{1+c \cdot \exp(-bt)}, \quad (5.29)$$

where $I(t)$ means the mean value functions for the inflection S-shaped SRGM, based on a nonhomogeneous Poisson process (NHPP) [11], a the expected number of latent faults for SRGM, and b the fault-detection rate per fault. Generally, the parameter c is defined as $\frac{(1-l)}{l}$. We define the parameter l as the mean value of network traffic density in this chapter.

By using Itô's formula [15, 16], we can obtain the integrated solution of Eq. (5.24) under the initial condition $C(0) = 0$ as follows [14]:

$$F_c(t) = \alpha \left[1 - \exp\left\{ -\int_0^t b(s)ds - p_*(t)\sigma_1\omega_1(t) - q_*(t)\sigma_2\omega_2(t) \right\} \right]. \quad (5.30)$$

Therefore, the cumulative number of detected faults of edge operation are obtained as follows:

$$F_c(t) = \alpha \left[1 - \frac{1+c}{1+c \cdot \exp(-bt)} \right.$$
$$\left. \cdot \exp\left\{ -bt - p_*(t)\sigma_1\omega_1(t) - q_*(t)\sigma_2\omega_2(t) \right\} \right]. \quad (5.31)$$

5.3 Cost Optimization Based on Cyclically Two Dimensional Stochastic Differential Equation Model

Similarly, considering the characteristics of big fault data and version-upgrade, it is interesting for the software developers to predict and estimate the time when we should stop operating in order to maintain a cloud OSS efficiently. We define the following cost parameters:

c_1: the operation cost per unit time,
c_2: the fixing cost per fault during the operation,
c_3: the maintenance cost per fault after the operation.

Then, the software cost based on the cyclic Wiener cases in the operation of cloud OSS can be formulated as:

$$C_{c1}(t) = c_1 t + c_2 F_c(t). \tag{5.32}$$

Also, the expected software maintenance cost based on two noisy cases after the maintenance of cloud software system is represented as follows:

$$C_{c2}(t) = c_3 \{\alpha - F_c(t)\}. \tag{5.33}$$

Consequently, from Eqs. (5.32) and (5.33), the total expected software maintenance cost based on two noisy cases is given by

$$C_{ct}(t) = C_{c1}(t) + C_{c2}(t). \tag{5.34}$$

The optimum maintenance time t^* is obtained by minimizing $C_{ct}(t)$ in Eq. (5.34). Moreover, we can derive the total software cost based on two dimensional Wiener processes as follows:

$$C_{ct}(t) = c_1 t + c_2 F_c(t) + c_3 \{\alpha - F_c(t)\}. \tag{5.35}$$

5.4 Numerical Examples

We focus on *OpenStack* Project [17] including several edge components. In this chapter, we show numerical examples by using the data sets as shown in Tables 2.1, 2.2, 2.3, 2.4, 2.5, 2.6, 2.7, 2.8, 2.9, 2.10, 2.11 and 2.12. on the assumption of the edge OSS service.

The estimated cumulative number of detected faults is shown in Fig. 5.2. Also, the estimated amount of cumulative number of remaining faults is shown in Fig. 5.3. Moreover, the estimated total software cost is shown in Fig. 5.4.

Considering the estimated cumulative number of detected faults, the value of y-axis is given as follows:

$$S_1(t) \equiv S_y(t) = \alpha \left[1 - \frac{1+c}{1 + c \cdot \exp(-bt)} \cdot \exp\left\{ -bt - p(t)\sigma_1 \omega_1(t) \right\} \right]. \tag{5.36}$$

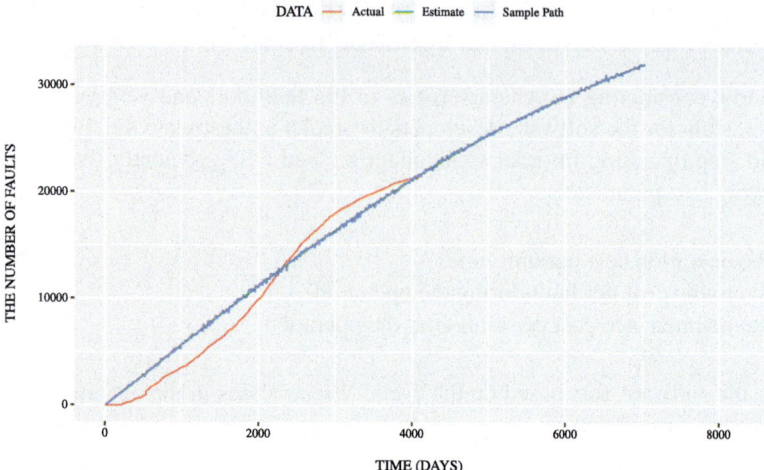

Fig. 5.2 The estimated cumulative number of detected faults

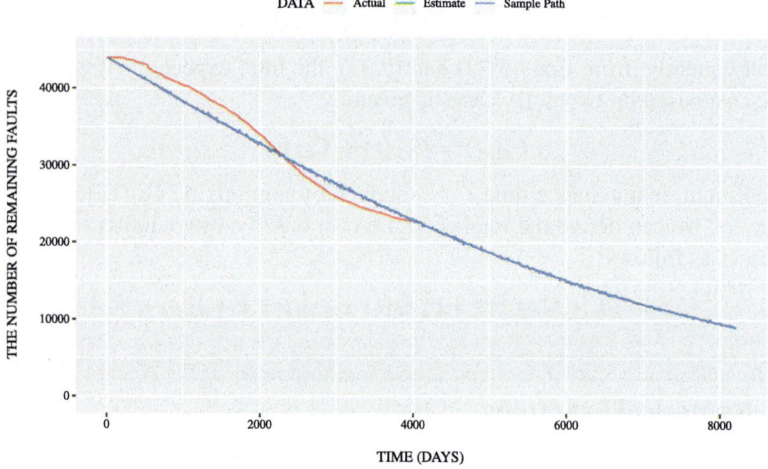

Fig. 5.3 The estimated cumulative number of remaining faults

Similarly, the value of z-axis is given as follows:

$$S_2(t) \equiv S_z(t) = \alpha \left[1 - \frac{1+c}{1+c \cdot \exp(-bt)} \cdot \exp\left\{ -bt - q(t)\sigma_2 \omega_2(t) \right\} \right]. \quad (5.37)$$

Then, the estimated cumulative number of detected faults in case of sawtooth wave, the estimated cumulative number of detected faults in case of square wave, and the estimated cumulative number of detected faults in case of mixed waves of sawtooth and square are respectively shown in Figs. 5.5, 5.6 and 5.7.

5.4 Numerical Examples

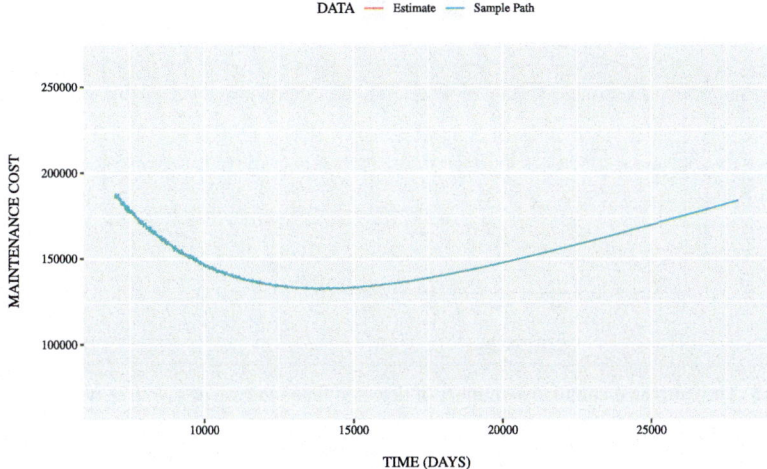

Fig. 5.4 The estimated total software cost

Also considering the estimated cumulative number of remaining faults, the value of y-axis is given as follows:

$$\overline{S_1}(t) \equiv \overline{S_y}(t) = \alpha \cdot \frac{1+c}{1+c \cdot \exp(-bt)} \cdot \exp\left\{-bt - p(t)\sigma_1\omega_1(t)\right\}. \quad (5.38)$$

Similarly, the value of z-axis is given as follows:

$$\overline{S_2}(t) \equiv \overline{S_z}(t) = \alpha \cdot \frac{1+c}{1+c \cdot \exp(-bt)} \cdot \exp\left\{-bt - q(t)\sigma_2\omega_2(t)\right\}. \quad (5.39)$$

Then, the estimated cumulative number of remaining faults in case of sawtooth wave, the estimated cumulative number of remaining faults in case of square wave, and the estimated cumulative number of remaining faults in case of mixed waves of sawtooth and square are respectively shown in Figs. 5.8, 5.9 and 5.10.

In case of $n = 7$, the estimated weight functions Eqs. (5.17) and (5.18) for Wiener processes $\omega_i(t)$ for edge operational time t is respectively shown in Figs. 5.11, 5.12 and 5.13 in cases of the sawtooth wave, the square wave, and the mixed waves of sawtooth and square.

Furthermore, Figs. 5.14, 5.15 and 5.16 show the estimated sample paths of Wiener process terms at edge operational time t for cases of the sawtooth wave, the square wave, and the mixed waves of sawtooth and square, respectively.

From above mentioned results, we have found that our model can describe the characteristics of edge computing according to the changes of the network environment. In particular, the weight functions are approaching each wave according to the increase of parameter n.

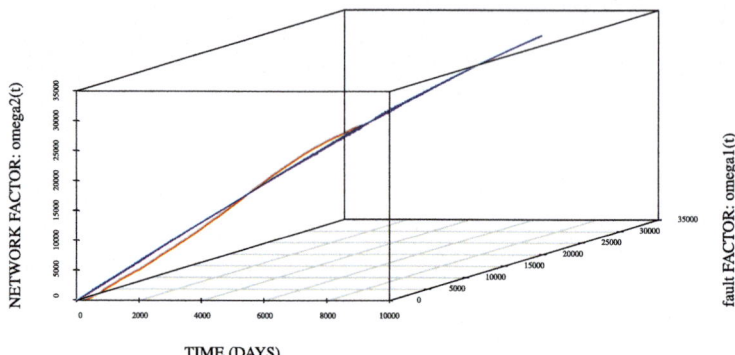

Fig. 5.5 The estimated cumulative number of detected faults in case of sawtooth wave

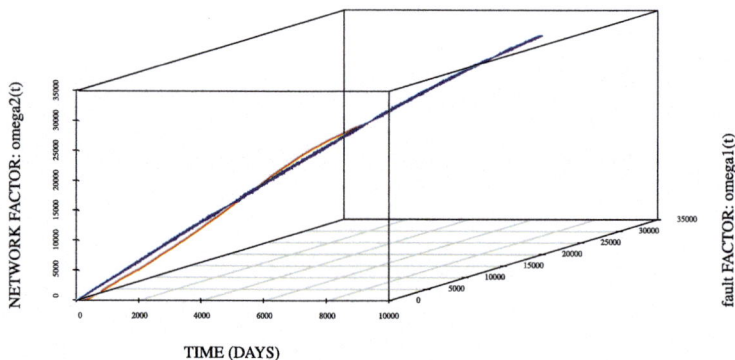

Fig. 5.6 The estimated cumulative number of detected faults in case of square wave

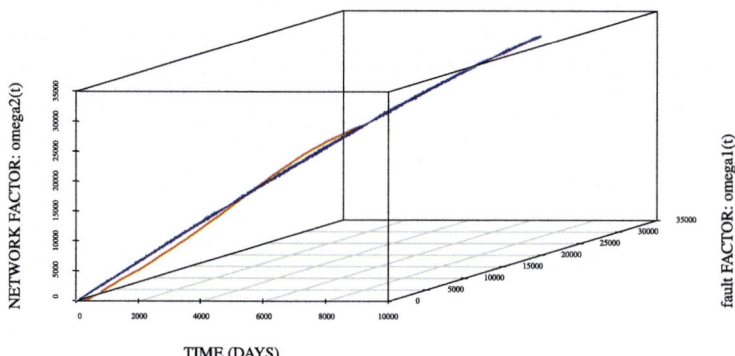

Fig. 5.7 The estimated cumulative number of detected faults in case of mixed waves of sawtooth and square

5.4 Numerical Examples

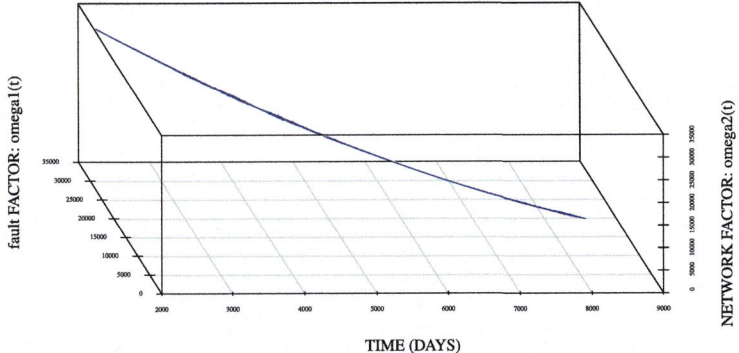

Fig. 5.8 The estimated cumulative number of remaining faults in case of sawtooth wave

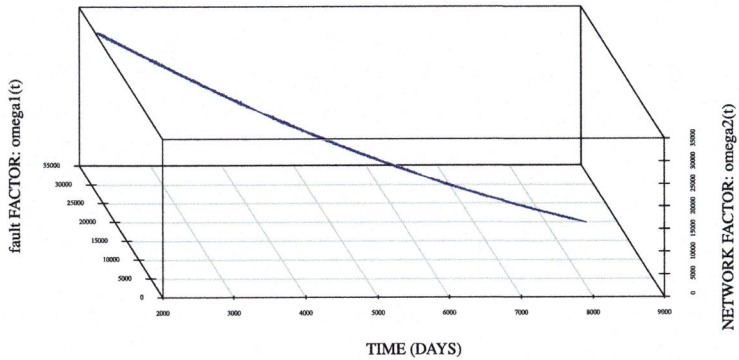

Fig. 5.9 The estimated cumulative number of remaining faults in case of square wave

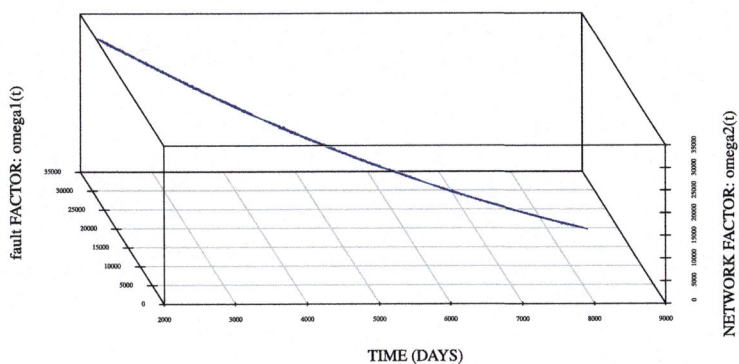

Fig. 5.10 The estimated cumulative number of remaining fault in case of mixed waves of sawtooth and square

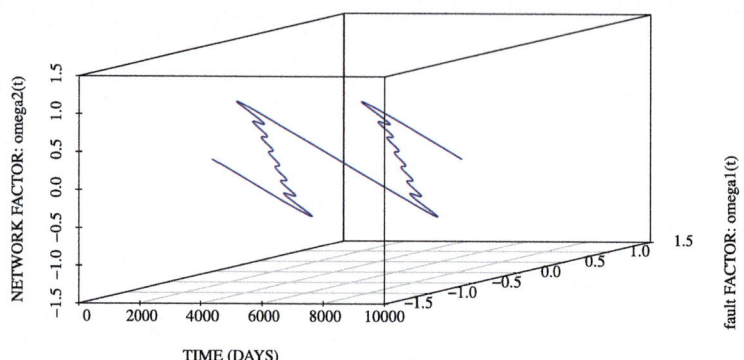

Fig. 5.11 The estimated weight functions $p_1(t)$ and $q_1(t)$ in case of sawtooth wave

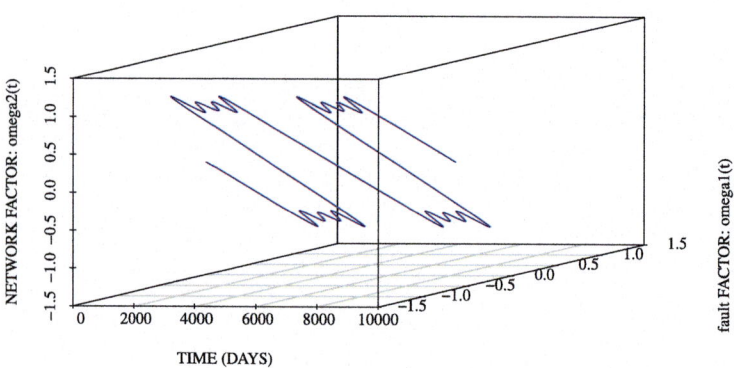

Fig. 5.12 The estimated weight functions $p_2(t)$ and $q_2(t)$ in case of square wave

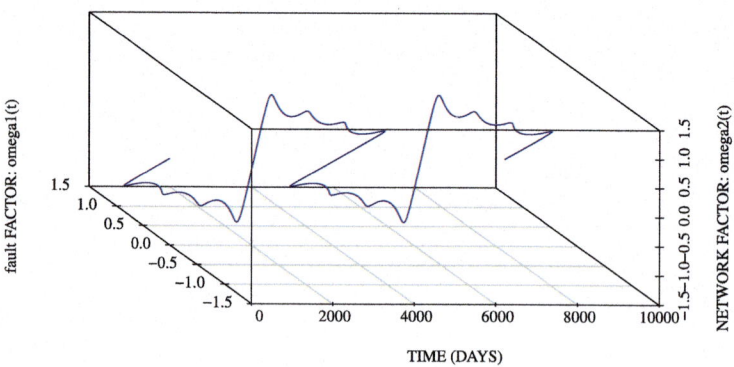

Fig. 5.13 The estimated weight functions $p_1(t)$ and $q_2(t)$ in case of mixed waves of sawtooth and square

5.4 Numerical Examples

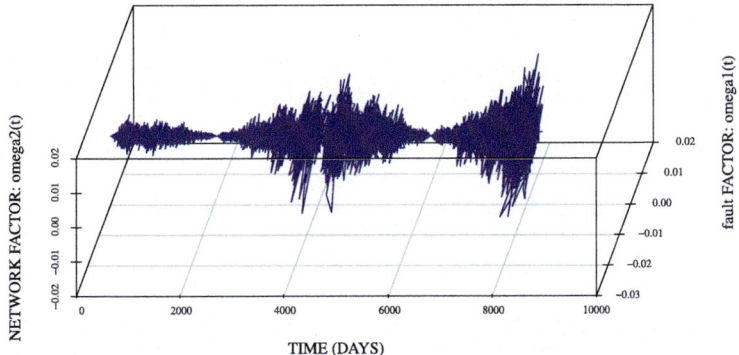

Fig. 5.14 The estimated sample paths of Wiener process terms at edge operational time t in case of sawtooth wave

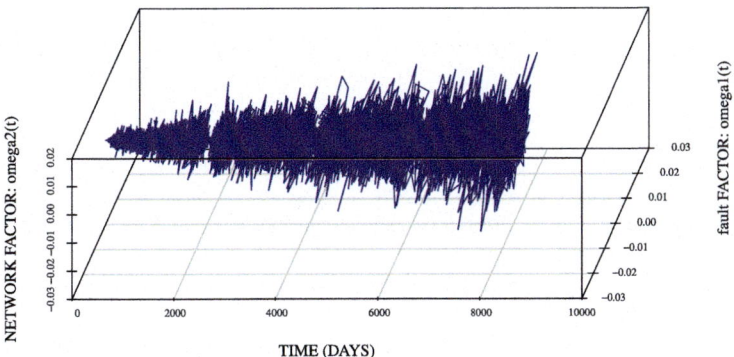

Fig. 5.15 The estimated sample paths of Wiener process terms at edge operational time t in case of square wave

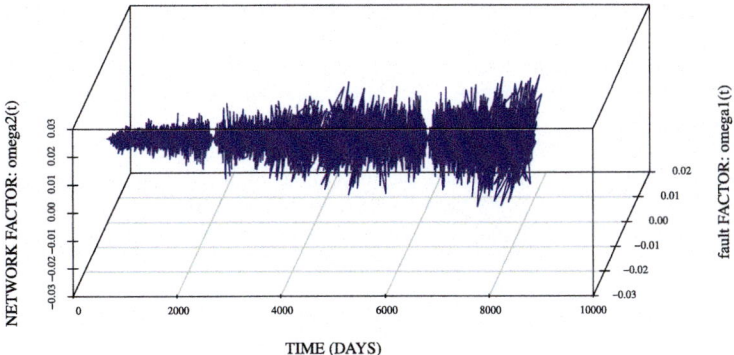

Fig. 5.16 The estimated sample paths of Wiener process terms at edge operational time t in case of mixed waves of sawtooth and square

5.5 Concluding Remarks

The appropriate fault control for the maintenance of edge software service will indirectly depends on the placement environment of edge server. This chapter has proposed the two dimensional Wiener process model with cyclic noisy function considering the irregular fluctuations as the characteristics of edge server. It is difficult for the edge server managers to control the progress of server maintenance. This chapter has discussed two dimensional Wiener process modeling with cyclic noisy function by using the actual edge OSS fault data as follows:

◎ Edge OSS effort expenditure with time-variation
◎ Two dimensional Wiener processes
◎ Cyclic changing of noises with time-dependent

The proposed model will be helpful as the assessment measures of the reliability control for edge OSS service in operation phase.

References

1. Ibrahim IM et al (2018) A robust generic multi-authority attributes management system for cloud storage services. IEEE Trans Cloud Comput. https://doi.org/10.1109/TCC.2018.2867871, 30 Aug 2018
2. Ahmad AA et al (2019) Scalability analysis comparisons of cloud-based software services. J Cloud Comput Adv Syst Appl. https://doi.org/10.1186/s13677-019-0134-y, 23 Jul 2019
3. Ozcan MO, Odaci F, Ari I (2019) Remote debugging for containerized applications in edge computing environments. In: Proceedings of the 2019 IEEE international conference on edge computing (EDGE), Milan, Italy, pp 30–32. https://doi.org/10.1109/EDGE.2019.00021
4. Ngoko Y, Cérin C (2017) An edge computing platform for the detection of acoustic events. In: Proceedings of the 2017 IEEE international conference on edge computing (EDGE), Honolulu, HI, USA, pp 240–243. https://doi.org/10.1109/IEEE.EDGE.2017.44
5. Caprolu M, Di Pietro R, Lombardi F, Raponi S (2019) Edge computing perspectives: architectures, technologies, and open security issues. In: Proceedings of the 2019 IEEE international conference on edge computing (EDGE), Milan, Italy, pp 116–123. https://doi.org/10.1109/EDGE.2019.00035
6. Dolui K, Datta SK (2017) Comparison of edge computing implementations: fog computing, cloudlet and mobile edge computing. In: Proceedings of the 2017 global internet of things summit (GIoTS), Geneva, Switzerland, pp 1–6. https://doi.org/10.1109/GIOTS.2017.8016213
7. Yamada S, Ohtera H, Narihisa H (1986) Software reliability growth models with testing-effort. IEEE Trans Reliab 35(1):19–23. https://doi.org/10.1109/TR.1986.4335332
8. Zhoum Y, Davis J (2005) Open source software reliability model: an empirical approach. In: Proceedings of the workshop on open source software engineering (WOSSE), vol 30, no 4, pp 67–72
9. Li P, Shaw M, Herbsleb J, Ray B, Santhanam P (2004) Empirical evaluation of defect projection models for widely-deployed production software systems. In: Proceeding of the 12th international symposium on the foundations of software engineering (FSE-12), pp 263–272
10. Norris J (2004) Mission-critical development with open source software. IEEE Softw Mag 21(1):42–49

References

11. Yamada S (2014) Software reliability modeling: fundamentals and applications. Springer, Tokyo/Heidelberg
12. Kapur PK, Pham H, Gupta A, Jha PC (2011) Software reliability assessment with OR applications. Springer, London
13. Tamura Y, Yamada S (2018) Multi-dimensional software tool for OSS project management considering cloud with big data. Int J Reliab Qual Saf Eng 25(3):1850014-1–1850014-16. World Scientific
14. Yamada S, Kimura M, Tanaka H, Osaki S (1994) Software reliability measurement and assessment with stochastic differential equations. IEICE Trans Fundam E77–A(1):109–116
15. Arnold L (1974) Stochastic differential equations-theory and applications. Wiley, New York
16. Wong E (1971) Stochastic processes in information and systems. McGraw-Hill, New York
17. The OpenStack project, Build the future of Open Infrastructure. https://www.openstack.org/

Cyclically Two Dimensional Jump Diffusion Process Modeling

6.1 Introduction

Many OSS projects are existing around the world. OSS are operated under the irregular status of component-collision. In particular, several components of OSS will be made a change according to the progress of OSS development. Also, OSS include several versions such as bug-fix version, minor version, and major version. Considering the characteristics of OSS projects, the operation performance of OSS development will take an irregular fluctuation during the operation, because several OSS components are updated with the OSS version-upgrade. In particular, the characteristics of OSS changes with the switch in several components according to several version upgrades. Several research papers in terms of the reliability of OSS have been proposed by several researchers [1–3]. In the past, the methods of software reliability assessment based on the SRGM's have been proposed by several researchers [4,5]. Also, our research group has been proposed the method of reliability assessment for the OSS [6].

This chapter proposes a method of OSS project management considering irregular fluctuation in reliability resulting from the characteristics with component dependency of OSS development and management. In particular, the OSS project assessment method based on the jump diffusion process model considering the component dependency in terms of the reliability is proposed in order to comprehend the component characteristics of OSS in this chapter.

6.2 Weighted Jump Diffusion Process Modeling

We apply a stochastic differential equation model to control the reliability in the operational phase of OSS projects. Then, let $\Psi(t)$ be the cumulative number of detected faults t $(t \geq 0)$ in the operation of OSS project. Suppose that $\Psi(t)$ takes on continuous real values. Since the estimated number of detected faults is observed during the operational phase of the OSS project, $\Psi(t)$ gradually increases as the operational procedures go on. Based on software reliability growth modeling [4], the following linear differential equation in terms of the reliability management can be formulated:

$$\frac{d\Psi(t)}{dt} = \beta(t)\{\alpha - \Psi(t)\}, \tag{6.1}$$

where $\beta(t)$ is the fault-detection rate at OSS operational time t and a non-negative function, α, means the number of latent faults for OSS.

Therefore, we extend Eq. (6.1) to the following stochastic differential equation with Brownian motion [7,8]:

$$\frac{d\Psi(t)}{dt} = \{\beta(t) + \sigma\nu(t)\}\{\alpha - \Psi(t)\}, \tag{6.2}$$

where σ is a positive constant representing a magnitude of the irregular fluctuation, and $\nu(t)$ a standardized Gaussian white noise. Then, we extend Eq. (6.2) to the following stochastic differential equation of an Itô type [7]:

$$d\Psi(t) = \left\{\beta(t) - \frac{1}{2}\sigma^2\right\}\{\alpha - \Psi(t)\}dt + \sigma\{\alpha - \Psi(t)\}d\omega(t), \tag{6.3}$$

where $\omega(t)$ is one-dimensional Wiener process which is formally defined as an integration of the white noise $\nu(t)$ with respect to time t.

We can added the jump term to the proposed stochastic differential equation models in order to represent the irregular state around the time t influenced by various external factors in the development phase of OSS project. Then, the jump-diffusion process [9] is given as follows.

$$d\Psi_j(t) = \left\{\beta(t) - \frac{1}{2}\sigma^2\right\}\{\alpha - \Psi_j(t)\}dt$$
$$+ \sigma\{\alpha - \Psi_j(t)\}d\omega(t)$$
$$+ d\left\{\sum_{i=1}^{N_t(\rho)}(J_i - 1)\right\}, \tag{6.4}$$

where $N_t(\rho)$ is a Poisson point process with parameter ρ at operation time t. Also, $N_t(\rho)$ and ρ are defined as the number of occurred jumps and ρ the jump rate, respectively. $N_t(\rho)$, $\omega(t)$, and J_i are assumed to be mutually independent [10,11]. Moreover, J_i represents ith jump range. Furthermore, we consider the weighted jump diffusion process model as follows:

$$d\Psi_j(t) = \left\{\beta(t) - \frac{1}{2}\sigma^2\right\}\{\alpha - \Psi_j(t)\}dt$$
$$+ p_*(t) \cdot \sigma\{\alpha - \Psi_j(t)\}d\omega(t)$$
$$+ q_*(t) \cdot d\left\{\sum_{i=1}^{N_t(\rho)}(J_i - 1)\right\}, \tag{6.5}$$

where $p(t)$ is the weight function for Wiener process in Eq. (5.17), $q(t)$ are the weight function for Wiener process in Eq. (5.18) at OSS operational time t.

Generally, the jump diffusion models have been applied to the area of option pricing [9]. In particular, it is difficult to directly apply the idea of existing option pricing model to the OSS fault-detection phenomena, because the log-normal distribution is optimized for the option pricing area. Also, it is unnatural to apply the log-normal distribution based on the option pricing model to the OSS fault-detection phenomena, because it is usually assumed that the OSS fault-detection phenomena have non-biased distribution in the research area of software reliability engineering. Therefore, we define the following normal distribution function as Gaussian Jump-diffusion process in order to consider the characteristics of OSS fault-detection phenomena:

$$J_i \equiv g_i(x) = \frac{1}{\sqrt{2\pi}\tau} \exp\left[-\frac{(x-\mu)^2}{2\tau^2}\right]. \tag{6.6}$$

Then, we assume that the ith jump range J_i are independently estimated as the positive values in almost all cases, because the mean value μ keep a large value.

By using Itô's formula [7, 8, 12], the solution of the former equation can be obtained as follows:

$$\Psi_j(t) = \alpha \left[1 - \exp\left\{-\int_0^t b(s)ds - p_*(t)\sigma\omega(t) - q_*(t) \sum_{i=1}^{N_t(\rho)} \log J_i\right\}\right]. \tag{6.7}$$

Similarly, the estimated cumulative number of remaining faults based on or model can give as follows:

$$\Psi_{rj}(t) = \alpha \cdot \exp\left\{-\int_0^t b(s)ds - p_*(t)\sigma\omega(t) - q_*(t) \sum_{i=1}^{N_t(\rho)} \log J_i\right\}. \tag{6.8}$$

6.3 Cost Optimization Based on Cyclically Two Dimensional Jump Diffusion Process Model

Similarly, considering the characteristics of big fault data and version-upgrade, it is interesting for the software developers to predict and estimate the time when we should stop operating in order to maintain a cloud-edge OSS efficiently. We define the following cost parameters:

c_1: the operation cost per unit time,
c_2: the fixing cost per fault during the operation,
c_3: the maintenance cost per fault after the operation.

Then, the software cost based on the cyclic jump cases in the operation of cloud OSS can be formulated as:

$$C_{tj1}(t) = c_1 t + c_2 \Psi_j(t). \tag{6.9}$$

Also, the expected software maintenance cost based on jump noisy cases after the maintenance of cloud-edge software system is represented as follows:

$$C_{tj2}(t) = c_3\{\alpha - \Psi_j(t)\}. \tag{6.10}$$

Consequently, from Eqs. (6.9) and (6.10), the total expected software maintenance cost based on jump cases is given by

$$C_{tj}(t) = C_{tj1}(t) + C_{tj2}(t). \tag{6.11}$$

The optimum maintenance time t^* is obtained by minimizing $C_{tj}(t)$ in Eq. (6.11). Moreover, we can derive the total software cost based on two dimensional Wiener processes as follows:

$$C_{tj}(t) = c_1 t + c_2 \Psi_j(t) + c_3\{\alpha - \Psi_j(t)\}. \tag{6.12}$$

6.4 Numerical Examples

We focus on *OpenStack* Project [13] including several edge components. In this chapter, we show numerical examples by using the data sets as shown in Tables 2.1, 2.2, 2.3, 2.4, 2.5, 2.6, 2.7, 2.8, 2.9, 2.10, 2.11 and 2.12 on the assumption of the edge OSS service.

The estimated cumulative number of detected faults is shown in Fig. 6.1. Also, the estimated weighted function is shown in Fig. 6.2. Moreover, the estimated noisy term is shown in Fig. 6.3. Furthermore, the estimated total software cost is shown in Fig. 6.4.

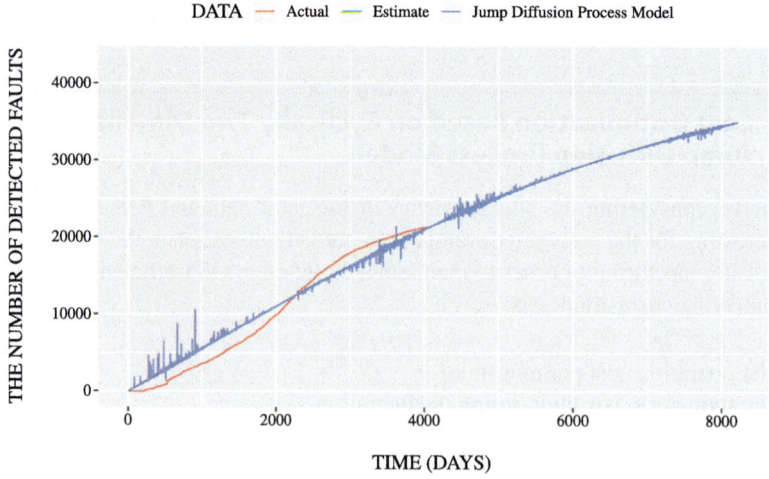

Fig. 6.1 The estimated cumulative number of detected faults

6.5 Concluding Remarks

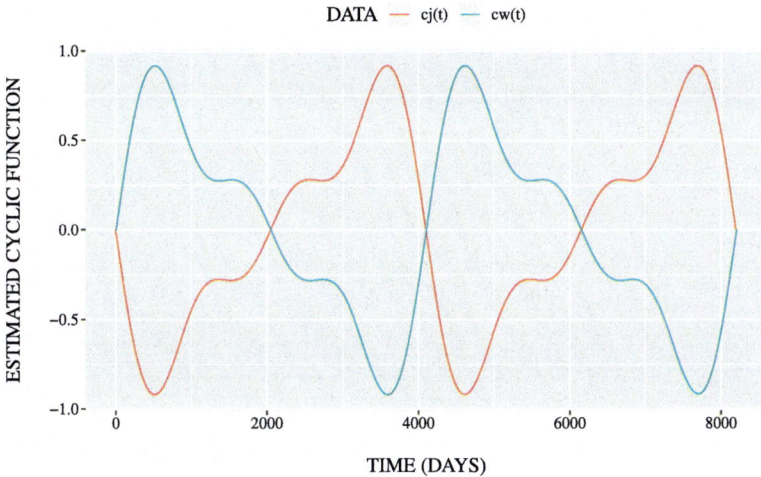

Fig. 6.2 The estimated weighted function

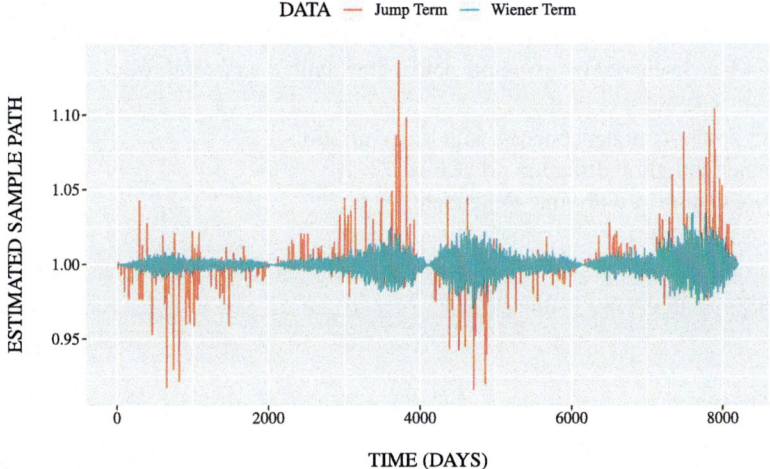

Fig. 6.3 The estimated noisy term

6.5 Concluding Remarks

The appropriate reliability control for OSS development will indirectly depends on the usage quality and cost reduction of OSS. This chapter has proposed a method of OSS project assessment considering the irregular fluctuations with jump and continuous noise from the non-contiguous characteristics of OSS development and

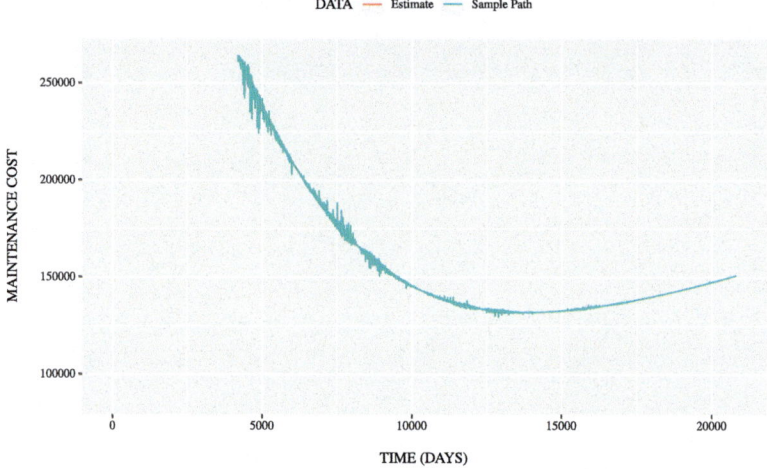

Fig. 6.4 The estimated total software cost

management. It is difficult for the OSS project managers to manage the progress of OSS development considering the range of the non-contiguous prediction. This chapter has discussed by using the actual OSS fault data as follows:

- OSS component dependency with time-variation
- Wiener and jump diffusion processes
- Noise changing with time-dependent.

The proposed method will be helpful as the method of the reliability assessment for OSS project in operation phase.

References

1. Zhoum Y, Davis J (2005) Open source software reliability model: an empirical approach. Proceed Workshop Open Source Softw Eng (WOSSE) 30(4):67–72
2. Li P, Shaw M, Herbsleb J, Ray B, Santhanam P (2004) Empirical evaluation of defect projection models for widely-deployed production software systems. In: Proceeding of the 12th international symposium on the foundations of software engineering (FSE–12), pp. 263–272
3. Norris J (2004) Mission-critical development with open source software. IEEE Softw Mag 21(1):42–49
4. Yamada S (2014) Software reliability modeling: fundamentals and applications. Springer-Verlag, Tokyo/Heidelberg
5. Kapur PK, Pham H, Gupta A, Jha PC (2011) Software reliability assessment with OR applications. Springer-Verlag, London
6. Yamada S, Tamura Y (2016) OSS reliability measurement and assessment, Springer series in reliability engineering. Springer

References

7. Arnold L (1974) Stochastic differential equations-theory and applications. John Wiley & Sons, New York
8. Wong E (1971) Stochastic processes in information and systems. McGraw-Hill, New York
9. Merton RC (1976) Option pricing when underlying stock returns are discontinuous. J Finan Econ 3(1–2):125–144
10. Tamura Y, Sone H, Yamada S (2020) Flexible jump diffusion process models for open source project with application to the optimal maintenance problem. In: International journal of reliability, quality and safety engineering, vol. 27, No. 6, World Scientific, pp. 2050020-1–2050020-18
11. Tamura Y, Sone H, Yamada S (2019) Productivity assessment based on jump diffusion model considering the effort management for OSS project. In: International journal of reliability, quality and safety engineering, vol. 26, No. 5, World Scientific, pp. 1950022-1–1950022-22
12. Yamada S, Kimura M, Tanaka H, Osaki S (1994) Software reliability measurement and assessment with stochastic differential equations. IEICE Trans Fundament E77–A(1):109–116
13. The OpenStack project, Build the future of Open Infrastructure, https://www.openstack.org/

Three Dimensional Tool Based on Noisy Model

7.1 Development of Prototype Tool

Our research group has proposed several reliability assessment tools. In particular, we have developed a three-dimensional tool for OSS reliability assessment. It is useful to easily understand the trend of reliability from various points of view by using three-dimensional modeling. We show the cumulative number of detected faults $M_*(t)$ at time t of our three-dimensional model proposed in the past as follows [1–3]:

$$M_1(t) = R_1(t)\left[1 - \frac{1+c}{1+c\cdot\exp(-bt)}\cdot\exp\left\{-bt - \sigma_1\omega_1(t)\right\}\right], \quad (7.1)$$

$$M_2(t) = R_2(t)\left[1 - \frac{1+c}{1+c\cdot\exp(-bt)}\cdot\exp\left\{-bt - \sigma_2\omega_2(t)\right\}\right], \quad (7.2)$$

$$M_3(t) = R_3(t)\left[1 - \frac{1+c}{1+c\cdot\exp(-bt)}\cdot\exp\left\{-bt - \sigma_3\omega_3(t)\right\}\right], \quad (7.3)$$

where $R_i(t)$ ($i = 1, 2, 3$) is the amount of changes in terms of specification according to each version of OSS. In addition, $R_i(t)$ ($i = 1, 2, 3$) is defined as $\alpha_i e^{-\beta_i t}$, where α_i ($i = 1, 2, 3$) is the number of latent faults in the OSS used in cloud computing and β_i ($i = 1, 2, 3$) is the changing rate in terms of specification according to each version of OSS. Then, we assume that the fault-prone specification for each version of OSS grows exponentially according to time t. On the other hand, the OSS will show a regression trend of reliability if β_i ($i = 1, 2, 3$) is a negative value. Conversely, the OSS will show a reliability growth trend if β_i ($i = 1, 2, 3$) is a positive value. Moreover, σ_1, σ_2, and σ_3 are noisy factors in terms of the magnitude of noisy fluctuation. $\omega_i(t)$ is the i-th Wiener process. Furthermore, b is the fault-detection rate per fault and c is defined as the parameter of fault factor.

In the proposed model, by considering the independence of each noise, we can assume that the parameter σ_1 means the failure-occurrence phenomenon due to inherent faults. The parameter σ_2 means the network changing rate per unit time resulting from OSS cloud computing. The parameter σ_3 means the renewal rate per unit time resulting from big data.

Considering our model in Eq. (6.7), we will be able to understand edge OSS reliability by using three-dimensional modeling. Moreover, we show the steps of the proposed prototype as follows:

1. The user of the prototype starts from the main menu by running our application. In addition, the user completes data preprocessing.
2. The user selects the calculation button. Then, the application window calls the Python program. In addition, the Python program imports the tensorflow package. Moreover, the proposed stochastic model is executed.
3. After the completion of the estimation of unknown parameters, several reliability assessment measures are illustrated by selecting the graph button.

We focus on *OpenStack* Project [4] including several edge components. In this chapter, we show numerical examples by using the fault data sets on the assumption of the edge OSS service. We apply the data in the above-mentioned from step 1 to step 3.

Considering the structure of the software tool, there are several visualization tools, such as the flowchart, activity diagram, and sequence diagram, in terms of UML. As reference, the flowchart, activity diagram, and sequence diagram of our tools in the past show in Figs. 7.1, 7.2 and 7.3. Generally, in many cases, UML has been used in the case where software is developed from scratch by a software manufacturer. However, our tool is implemented by using the package-based development style. This is the characteristic of our research. In the case of package-based development, we can show the structure of our prototype by using the package diagram in UML. The reasons are as follows:

- Different programming languages (HTML, CSS, JavaScript, and Python).
- Dynamic links based on Node.js and the file system of OS, such as macOS, Windows, Linux, etc.
- Difference in the scale and language of the packages.

From the above-mentioned characteristics, we show the structure of our tool by using the package diagram. Then, Fig. 7.4 shows the package diagram of the prototype developed by using UML.

Our tool is developed as a prototype. We believe that the task of researchers only provides until the point of a prototype. In addition, researchers should propose the application framework of the proposed method. Therefore, we show the framework based on the proposed method in this chapter. The completed tool will be able to be

7.2 Performance Illustrations of the Developed 3D Application

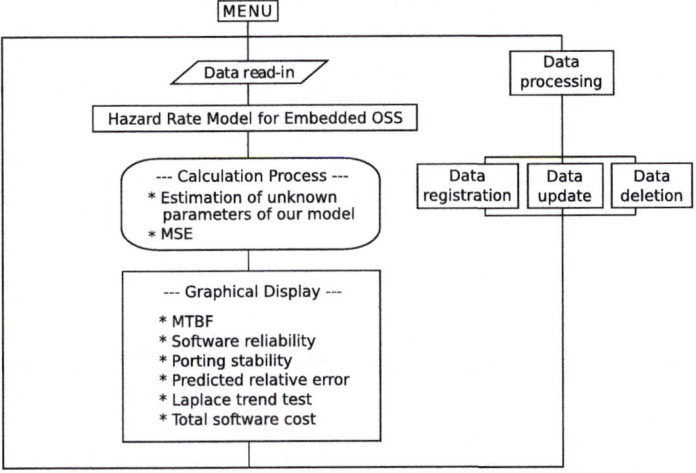

Fig. 7.1 The structure of our reliability/portability assessment tool for embedded OSS in the past

easily developed by software developers and business people. Our prototype will be helpful for users and developers to assess the reliability in the operation of the edge OSS service.

7.2 Performance Illustrations of the Developed 3D Application

7.2.1 Data Set for Edge OSS Computing

We focus on the *OpenStack* Project [4] included several edge components. In this chapter, we show numerical examples by using data sets on the assumption of the edge OSS service. The data used in this chapter are collected from the bug-tracking system as shown in Tables 2.1, 2.2, 2.3, 2.4, 2.5, 2.6, 2.7, 2.8, 2.9, 2.10, 2.11 and 2.12.

The demonstration of our prototype tool is available from "DEMO APPLICATION" at the following URL; however, the function of calculation cannot execute considering the security:
http://www.tam.eee.yamaguchi-u.ac.jp/, accessed on 23 December 2023.

Our prototype tool has been released as the OSS based on GNU General Public License (GPL) in December 2023. The source code of our tool is available from "SOFTWARE" at the following URL:
http://www.tam.eee.yamaguchi-u.ac.jp/, accessed on 23 December 2023.

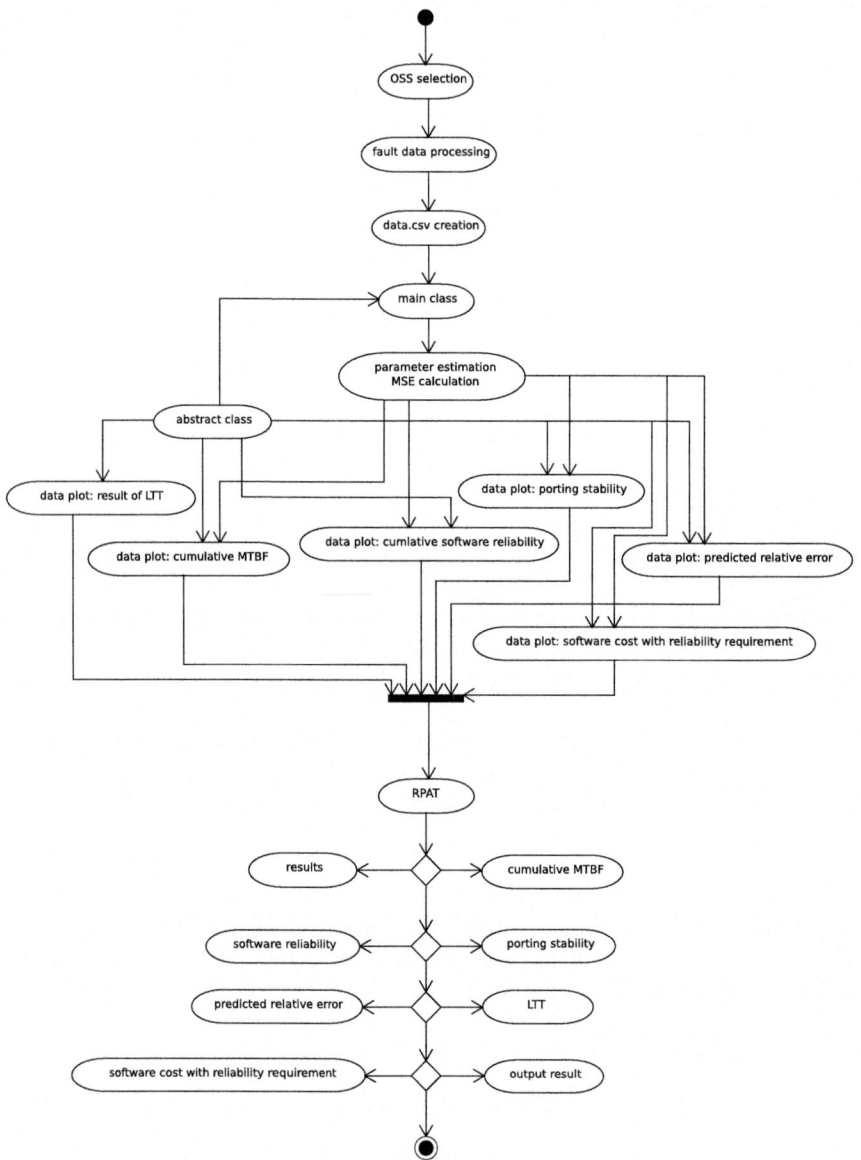

Fig. 7.2 The activity diagram of our reliability/portability assessment tool for embedded OSS in the past

7.2 Performance Illustrations of the Developed 3D Application

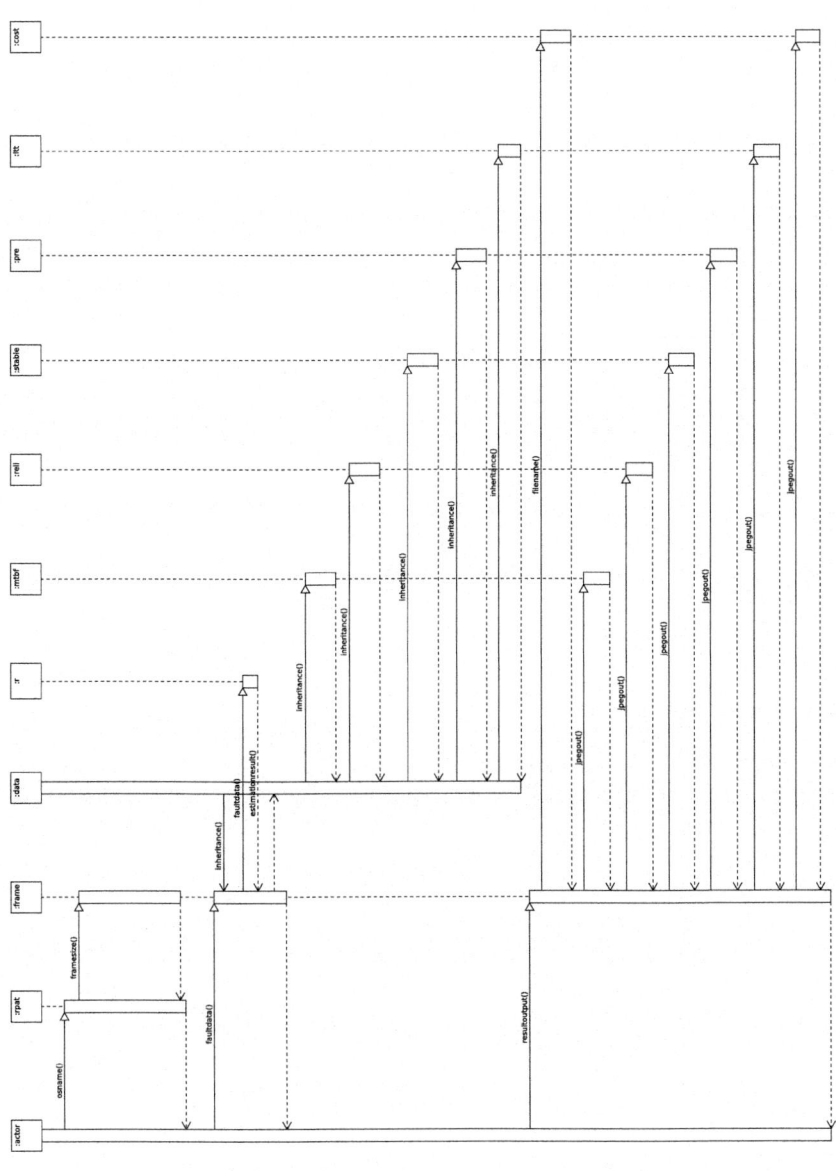

Fig. 7.3 The sequence diagram of our reliability/portability assessment tool for embedded OSS in the past

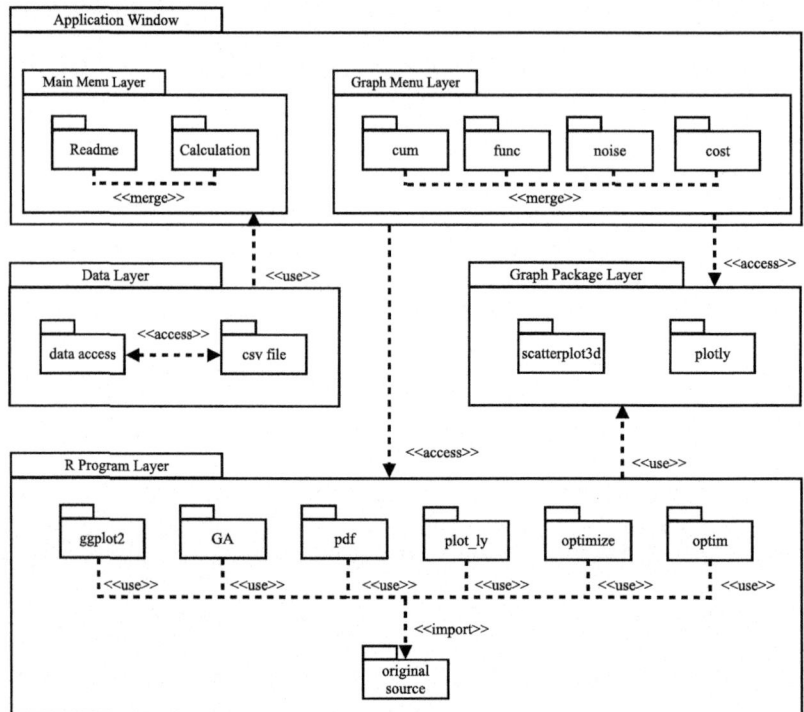

Fig. 7.4 The package diagram

7.2.2 Performance Results

First, we show the main screen of our tool in Fig. 7.5. In addition, Fig. 7.6 shows the menu of our tool. Moreover, Fig. 7.7 is the simplified readme screen of our tool. Our tool is structured by using the dynamic link based on NW.js and R. Therefore, we can program the simple menu by using the HTML and javascript codes.

Figures 7.8, 7.9, 7.10, 7.11 and 7.12 show the estimated cumulative number of detected faults, the estimated cumulative number of remaining faults, the estimated noise term, the estimated weighted function, and the estimated total software cost, respectively discussed in Chap. 2.

As for the above-mentioned results, we confirm that the developed prototype tool for OSS reliability assessment based on stochastic model is useful to estimate the reliability. In particular, the advantage of our method is that it can make use of the noisy cases.

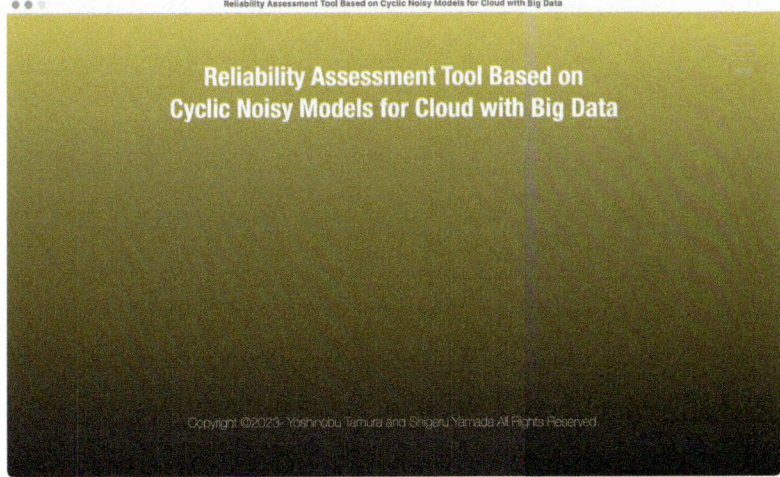

Fig. 7.5 The main screen of our tool

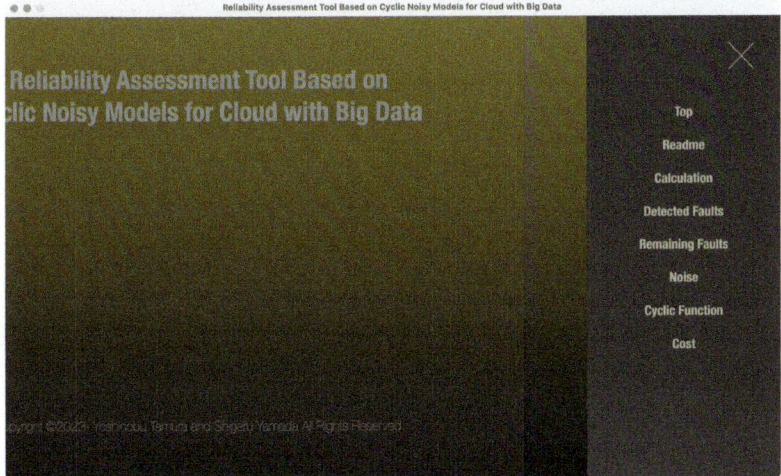

Fig. 7.6 The menu of our tool

7.3 Concluding Remarks

In the operation of the cloud service, several edge OSS components are embedded in cloud OSS computing. In this case, there is several noisy factors in the operation phase of edge computing.

In this chapter, we have developed the prototype tool by using three dimensional tool based on the noisy model.

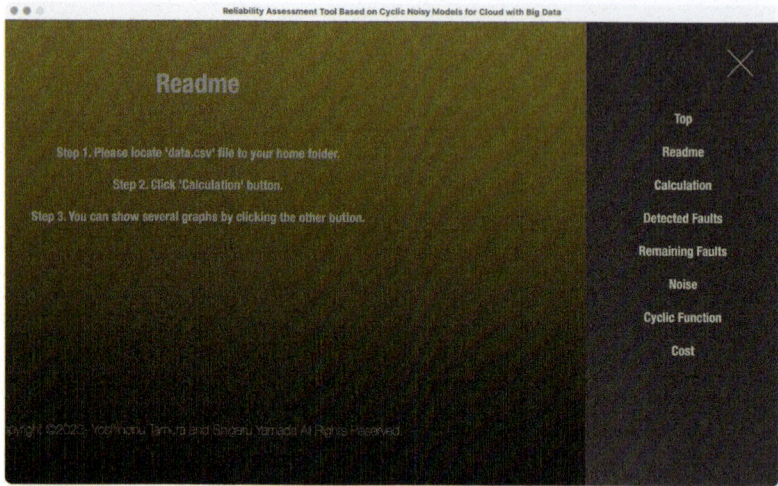

Fig. 7.7 The readme of our tool

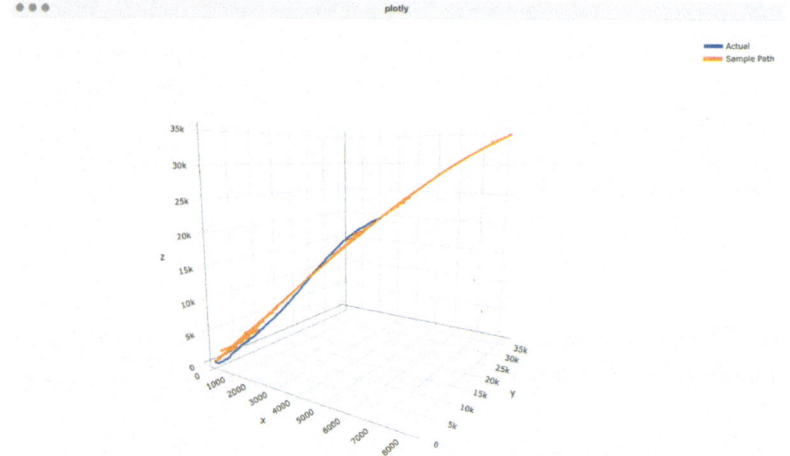

Fig. 7.8 The estimated cumulative number of detected fault

Finally, this chapter has discussed the reliability trend of OSS faults for edge computing. In addition, we have developed the prototype of a software tool based on the proposed method by using actual edge OSS data as follows:

- The cumulative number of detected fault as data preprocessing.
- The use of three dimensional tool by using noisy model based on the stochastic model.
- The development of a prototype as a reliability assessment tool.

7.3 Concluding Remarks

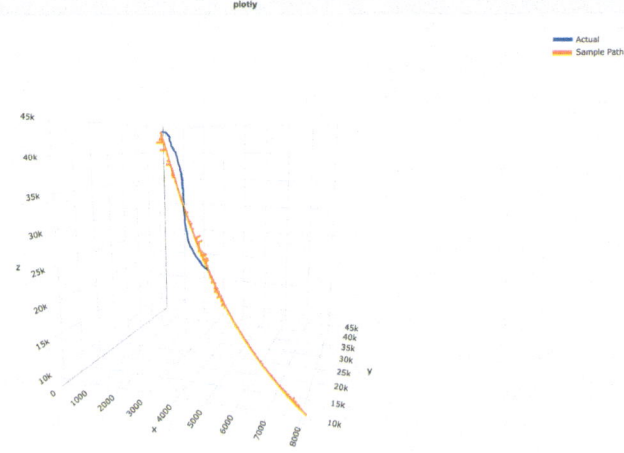

Fig. 7.9 The estimated cumulative number of remaining fault

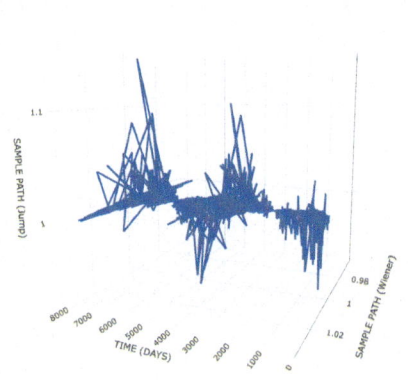

Fig. 7.10 The estimated noise term

The proposed method and prototype will be helpful as assessment measures of reliability control for an edge OSS service in the operation phase.

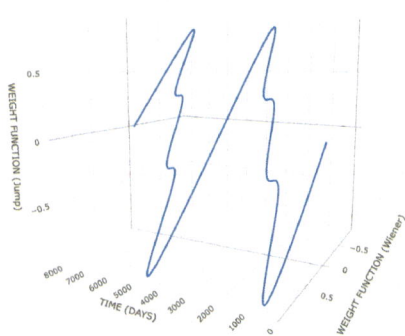

Fig. 7.11 The estimated weighted function

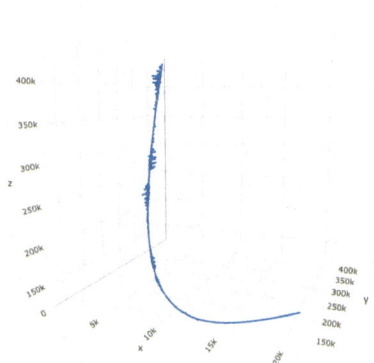

Fig. 7.12 The estimated total software cost

References

1. Tamura Y, Yamada S (2017) Dependability analysis tool considering the optimal data partitioning in a mobile cloud. In: Reliability modeling with computer and maintenance applications. World Scientific, pp. 45–60
2. Tamura Y, Yamada S (2017) Dependability analysis tool based on multi-dimensional stochastic noisy model for cloud computing with big data. Int J Math Eng Manage Sci 2(4):273–287

References

3. Tamura Y, Yamada S (2022) Prototype of 3D reliability assessment tool based on deep learning for edge OSS computing. In: Mathematics, vol. 10, No. 9, Multidisciplinary Digital Publishing Institute, Switzerland, https://doi.org/10.3390/math10091572, pp. 1–12
4. The OpenStack project, Build the future of Open Infrastructure, https://www.openstack.org/

Deep Learning Method Based on Fault Big Data Analysis for OSS Reliability Assessment

8.1 Introduction

Many OSS are useful for many software engineers and general users around the world. However, there is no established standard method of quality/reliability assessment for OSS. The bug tracking system is well known as the useful system for quality improvement of OSS. The bug tracking system is implemented in various open source projects in recent years. Also, various fault data sets are registered on the database of bug tracking system.

In the past, various SRGM's [1,2] have been applied to analyze the reliability for quality management of the system testing phase in software development. Also, several software reliability models for OSS have been proposed in several research papers [3]. However, it will be difficult for the software developers to select the optimal SRGM for the testing-phase of an actual software project. Moreover, the software managers need to prepare the fault count data in case of the software reliability models, i.e., it will be need for the software managers to convert the raw data of software faults recorded on the bug tracking system to the software fault count data. Then, it may take several trouble with the modification of fault count data sets from the raw data. Therefore, it will be difficult for the software developers to analyze the reliability/quality of OSS by using the software reliability models for OSS. On the other hand, it will be able to take a prompt action for quality management if the software developers can use the raw data recorded on the bug tracking system.

We discuss the recognition method of severity level for software fault in this chapter. Then, the method of big fault data analysis by using deep learning is proposed. Also, this chapter presents several numerical examples of the proposed method by using the actual big fault data. Moreover, the proposed method by using the deep learning is compared with the method by using neural network. Then, the proposed method is applied to the amount of learning data in order to assess the effectiveness of deep learning [4].

8.2 Big Fault Data Analysis by Using Neural Network

Considering general method of neural network by using back-propagation, we can give the input-output rules of each unit on each layer as follows:

$$H_j = f\left(\sum_{i=1}^{I} \omega_{ij}^1 I_i\right), \qquad (8.1)$$

$$R_k = f\left(\sum_{j=1}^{J} \omega_{jk}^2 H_j\right), \qquad (8.2)$$

where H_j is the activation function of hidden layer. Similarly, R_k is the activation function of response layer. Also, the following logistic function $f(\cdot)$ is widely-known as a sigmoid function.

$$f(x) = \frac{1}{1 + e^{-\theta x}}, \qquad (8.3)$$

where θ is the gain. The multi-layered neural networks [5] based on back-propagation in order to learn the fault severity levels from the big fault data recorded on bug tracking system is used in this chapter. The error function is given as follows:

$$E = \frac{1}{2} \sum_{k=1}^{K} (R_k - L_k)^2, \qquad (8.4)$$

where $L_k (k = 1, 2, \ldots, K)$ are the target input values as the output values.

In terms of the amount of bug tracking system, The following amount of information is used as parameters to the input data $I_i (i = 1, 2, \ldots, I)$.

---------- Input data used in neural network. ----------

- ▶ Recorded date
- ▶ Nickname of assignee
- ▶ Nickname of reporter
- ▶ Fault Status
- ▶ Operating system
- ▶ Software component name
- ▶ Software product name
- ▶ Software version name

8 kinds of fault level are selected as the amount of compressed characteristics, i.e., Critical, Major, Blocker, Regression, Normal, Minor, Enhancement, and Trivial, respectively. Then, the number of units for the response layer, K, is set to 8, because the number of fault levels is 8.

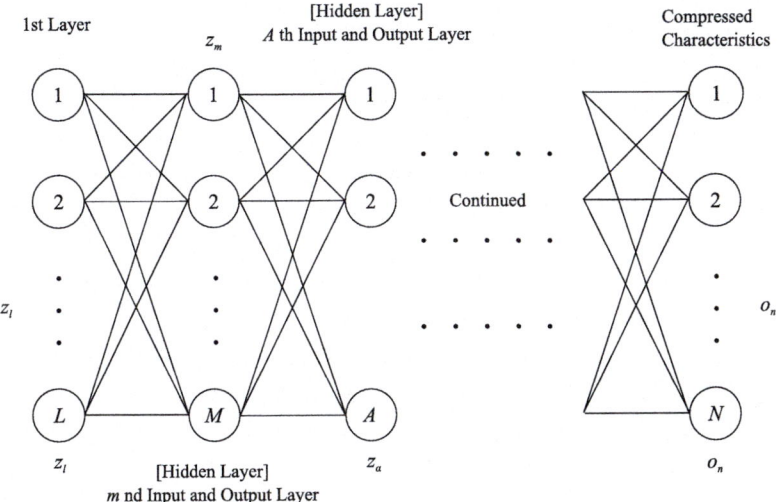

Fig. 8.1 The flow of deep learning

8.3 Big Fault Data Analysis by Using Deep Learning

The general flow of the deep learning is shown in Fig. 8.1. In Fig. 8.1, $Z_l(l = 1, 2, \ldots, L)$, $Z_\alpha(\alpha = 1, 2, \ldots, A)$, and $Z_m(m = 1, 2, \ldots, M)$ are the pre-training units. Moreover, the amount of compressed characteristics is shown as $O_n(n = 1, 2, \ldots, N)$. In terms of deep learning, several algorithms have been proposed [6–11] by several researchers in the past. Also, several software reliability assessment methods based on deep learning have been proposed [12, 13]. In order to learn the fault data registered on bug tracking system of open source software projects, the deep neural network is applied in this chapter.

Then, the fault levels as represented in Table 8.1 are used as the objective variable. Therefore, the amount of compressed characteristics is 8 kinds of fault level, i.e., Critical, Major, Blocker, Regression, Normal, Minor, Enhancement, and Trivial, respectively. As is the case in neural network, the following input data is applied as the explanatory variable to the input data $Z_l(l = 1, 2, \ldots, L)$.

───────────── Input data as the explanatory variable. ─────────────

- ▶ Recorded date
- ▶ Nickname of assignee
- ▶ Nickname of reporter
- ▶ Fault Status
- ▶ Operating system
- ▶ Software component name
- ▶ Software product name
- ▶ Software version name

Table 8.1 The indexed fault levels

Index number	Software fault level
1	Trivial
2	Enhancement
3	Minor
4	Normal
5	Regression
6	Blocker
7	Major
8	Critical

Then, these 8 kinds of explanatory variables means the amount of characteristics for pre-training units. In order to assess by using deep learning, the character data included in each input data as the explanatory variable is converted to the numerical value such as the occurrence ratio. Our research group has proposed the method of OSS reliability assessment based on above mentioned approach in the past [4].

On the other hand, as the other approach, we have proposed the method of reliability assessment as the case study of the detection time of software fault for the objective variable. Then, the input data sets are 13 kinds of factors from the bug tracking system.

8.4 Numerical Examples

As an example, we show the estimation results of the detection time of software fault by using the deep learning. We focus on the *OpenStack* Project [14] that included several edge components. Table 8.2 is a part of data set. All lines of data is over 20,000 lines. Then, the all data is over 260,000 data items. Actually, the users can obtain a greater number of data according to various OSS projects. In this chapter, we show numerical examples by using a part of data set as shown in Table 8.3 on the assumption of the edge OSS service.

The data used in this chapter are collected from the bug-tracking system. Several performance examples based on the data sets for OpenStack as OSS are shown in this chapter. In this chapter, the data recorded on the website of the bug tracking system in *OpenStack* Project. Then, 21,562 fault data set are obtained from website.

8.4 Numerical Examples

Table 8.2 A part of raw data logged on the bug tracking system

Bug ID	Opened	Changed	Reporter	Assignee	Product	Component	Status	Hardware	OS	Severity	Version
833398	2012/6/19 0:00	2019-09-09T16:04:58Z	David Naori	Xavier Queralt	Red Hat OpenStack	Openstack-nova	CLOSED	Unspecified	Unspecified	Low	1.0 (Essex)
835466	2012/6/26 0:00	2013-07-04T02:47:20Z	David Naori	Alan Pevec	Red Hat OpenStack	Openstack-nova	CLOSED	x86_64	Linux	High	1.0 (Essex)
842136	2012/7/22 0:00	2019-09-09T16:10:30Z	David Naori	Brent Eagles	Red Hat OpenStack	Openstack-nova	CLOSED	Unspecified	Unspecified	Medium	1.0 (Essex)
856263	2012/9/11 0:00	2022-07-09T07:12:40Z	Daniel Berrang	Daniel Berrang	Red Hat OpenStack	Openstack-nova	CLOSED	Unspecified	Unspecified	Unspecified	2.0 (Folsom)
857256	2012/9/13 0:00	2013-10-31T17:03:22Z	Russell Bryant	Alan Pevec	Red Hat OpenStack	Openstack-nova	CLOSED	Unspecified	Unspecified	High	1.0 (Essex)
857290	2012/9/14 0:00	2022-07-09T07:13:48Z	Kai(Kimi) Zhang	Alan Pevec	Red Hat OpenStack	Openstack-keystone	CLOSED	x86_64	Linux	High	2.0 (Folsom)
861175	2012/9/27 0:00	2016-04-22T05:01:41Z	Dan Yocum	Nikola Dipanov	Red Hat OpenStack	Openstack-nova	CLOSED	All	Linux	Medium	1.0 (Essex)
861492	2012/9/28 0:00	2016-04-22T05:01:41Z	Dan Yocum	Nikola Dipanov	Red Hat OpenStack	Openstack-nova	CLOSED	x86_64	Linux	Unspecified	1.0 (Essex)
861508	2012/9/28 0:00	2016-04-26T21:26:41Z	Dan Yocum	RHOS Maint	Red Hat OpenStack	Openstack-nova	CLOSED	Unspecified	Unspecified	Medium	1.0 (Essex)
861516	2012/9/28 0:00	2013-07-04T02:51:13Z	Dan Yocum	RHOS Maint	Red Hat OpenStack	Openstack-nova	CLOSED	Unspecified	Unspecified	Unspecified	1.0 (Essex)
861943	2012/10/1 0:00	2016-04-26T22:41:17Z	Attila Fazekas	Adam Young	Red Hat OpenStack	Openstack-keystone	CLOSED	Unspecified	Linux	Low	1.0 (Essex)
862322	2012/10/2 0:00	2019-09-10T14:14:01Z	Dan Yocum	Matthias Runge	Red Hat OpenStack	Openstack-keystone	CLOSED	Unspecified	Unspecified	Unspecified	1.0 (Essex)
865314	2012/10/11 0:00	2019-09-09T17:17:20Z	Yaniv Kaul	Sahid Ferdjaoui	Red Hat OpenStack	Openstack-nova	CLOSED	Unspecified	Unspecified	Medium	Unspecified

(continued)

Table 8.2 (continued)

Bug ID	Opened	Changed	Reporter	Assignee	Product	Component	Status	Hardware	OS	Severity	Version
865316	2012/10/11 0:00	2019-09-09T13:13:27Z	Yaniv Kaul	Nikola Dipanov	Red Hat OpenStack	Openstack-nova	CLOSED	Unspecified	Unspecified	Medium	1.0 (Essex)
865335	2012/10/11 0:00	2019-12-06T10:37:11Z	Yaniv Kaul	Matthew Booth	Red Hat OpenStack	Openstack-nova	CLOSED	Unspecified	Unspecified	Low	17.0 (Wallaby)
865336	2012/10/11 0:00	2016-04-27T00:21:03Z	Yaniv Kaul	Solly Ross	Red Hat OpenStack	Openstack-nova	CLOSED	Unspecified	Unspecified	High	1.0 (Essex)
865337	2012/10/11 0:00	2019-11-18T12:50:19Z	Yaniv Kaul	Daniel Berrang	Red Hat OpenStack	Openstack-nova	CLOSED	Unspecified	Unspecified	Unspecified	4
865338	2012/10/11 0:00	2016-04-22T05:01:41Z	Yaniv Kaul	Nikola Dipanov	Red Hat OpenStack	Openstack-nova	CLOSED	Unspecified	Unspecified	Medium	Unspecified
865340	2012/10/11 0:00	2013-07-04T02:52:27Z	Yaniv Kaul	Daniel Berrang	Red Hat OpenStack	Openstack-nova	CLOSED	Unspecified	Unspecified	Medium	1.0 (Essex)
865347	2012/10/11 0:00	2022-07-09T06:23:33Z	Yaniv Kaul	Martin Magr	Red Hat OpenStack	Openstack-packstack	CLOSED	Unspecified	Unspecified	High	1.0 (Essex)
865505	2012/10/11 0:00	2016-04-26T18:19:26Z	Yaniv Kaul	Daniel Berrang	Red Hat OpenStack	Openstack-nova	CLOSED	Unspecified	Unspecified	Medium	1.0 (Essex)
865924	2012/10/12 0:00	2013-02-06T20:21:23Z	Dan Yocum	RHOS Maint	Red Hat OpenStack	Openstack-nova	CLOSED	Unspecified	Unspecified	Unspecified	1.0 (Essex)
866193	2012/10/14 0:00	2013-07-04T02:52:28Z	Yaniv Kaul	RHOS Maint	Red Hat OpenStack	Openstack-nova	CLOSED	Unspecified	Unspecified	Medium	1.0 (Essex)
866199	2012/10/14 0:00	2016-04-22T05:01:41Z	Yaniv Kaul	Nikola Dipanov	Red Hat OpenStack	Openstack-nova	CLOSED	Unspecified	Unspecified	Medium	2.0 (Folsom)
866451	2012/10/15 0:00	2016-04-26T22:28:02Z	Jaroslav Henner	Alan Pevec	Red Hat OpenStack	Openstack-keystone	CLOSED	x86_64	Linux	Low	1.0 (Essex)
866681	2012/10/15 0:00	2013-07-04T02:52:28Z	Jaroslav Henner	RHOS Maint	Red Hat OpenStack	Openstack-keystone	CLOSED	x86_64	Linux	Medium	1.0 (Essex)

(continued)

8.4 Numerical Examples

Table 8.2 (continued)

Bug ID	Opened	Changed	Reporter	Assignee	Product	Component	Status	Hardware	OS	Severity	Version
867029	2012/10/16 0:00	2022-07-09T05:52:27Z	Alan Pevec	Alan Pevec	Red Hat OpenStack	Openstack-keystone	CLOSED	Unspecified	Unspecified	Unspecified	1.0 (Essex)
867035	2012/10/16 0:00	2016-04-22T05:01:41Z	Alan Pevec	Nikola Dipanov	Red Hat OpenStack	Openstack-nova	CLOSED	Unspecified	Unspecified	Unspecified	1.0 (Essex)
874316	2012/11/7 0:00	2013-07-04T02:53:38Z	Jaroslav Henner	Alan Pevec	Red Hat OpenStack	Openstack-nova	CLOSED	x86_64	Linux	Urgent	2.0 (Folsom)
875938	2012/11/12 0:00	2016-04-22T05:01:41Z	Jaroslav Henner	Nikola Dipanov	Red Hat OpenStack	Openstack-nova	CLOSED	x86_64	Linux	Medium	2.0 (Folsom)
877343	2012/11/16 0:00	2019-09-09T14:10:48Z	Jaroslav Henner	Pete Zaitcev	Red Hat OpenStack	Openstack-swift	CLOSED	x86_64	Linux	Low	2.0 (Folsom)
878229	2012/11/19 0:00	2016-04-27T02:40:37Z	Dan Yocum	Flavio Percoco	Red Hat OpenStack	Openstack-glance	CLOSED	All	Linux	Low	1.0 (Essex)
880980	2012/11/28 0:00	2013-07-04T02:54:47Z	Sandro Mathys	RHOS Maint	Red Hat OpenStack	Openstack-keystone	CLOSED	Unspecified	Unspecified	Unspecified	2.0 (Folsom)
880984	2012/11/28 0:00	2013-07-04T02:54:47Z	Nir Magnezi	Alan Pevec	Red Hat OpenStack	Openstack-keystone	CLOSED	Unspecified	Linux	Medium	2.0 (Folsom)
881048	2012/11/28 0:00	2013-01-29T11:09:50Z	Nir Magnezi	Martin Magr	Red Hat OpenStack	Openstack-glance	CLOSED	Unspecified	Linux	High	2.0 (Folsom)
881105	2012/11/28 0:00	2013-07-04T02:54:47Z	Jaroslav Henner	Adam Young	Red Hat OpenStack	Openstack-keystone	CLOSED	x86_64	Linux	Low	2.0 (Folsom)
881610	2012/11/29 0:00	2019-09-10T14:08:25Z	Nir Magnezi	Eric Harney	Red Hat OpenStack	Openstack-cinder	CLOSED	Unspecified	Linux	Medium	2.0 (Folsom)
881737	2012/11/29 0:00	2016-01-04T14:41:24Z	Pdraig Brady	Martin Magr	Red Hat OpenStack	Openstack-nova	CLOSED	Unspecified	Unspecified	Unspecified	1.0 (Essex)

(continued)

Table 8.2 (continued)

Bug ID	Opened	Changed	Reporter	Assignee	Product	Component	Status	Hardware	OS	Severity	Version
881740	2012/11/29 0:00	2016-04-26T13:55:02Z	Nir Magnezi	Eric Harney	Red Hat OpenStack	Openstack-cinder	CLOSED	Unspecified	Linux	High	2.0 (Folsom)
881810	2012/11/29 0:00	2022-07-09T06:05:38Z	Nir Magnezi	Martin Magr	Red Hat OpenStack	Openstack-nova	CLOSED	Unspecified	Linux	High	2.0 (Folsom)
882977	2012/12/3 0:00	2021-03-09T19:07:50Z	Nir Magnezi	Martin Magr	Red Hat OpenStack	Openstack-swift	CLOSED	Unspecified	Unspecified	Unspecified	2.0 (Folsom)
883829	2012/12/5 0:00	2022-07-09T05:53:06Z	Alan Pevec	Alan Pevec	Red Hat OpenStack	Openstack-keystone	CLOSED	Unspecified	Unspecified	Unspecified	2.0 (Folsom)
883831	2012/12/5 0:00	2012-12-11T13:08:13Z	Alan Pevec	Alan Pevec	Red Hat OpenStack	Openstack-glance	CLOSED	Unspecified	Unspecified	Unspecified	2.0 (Folsom)
883836	2012/12/5 0:00	2012-12-11T13:08:16Z	Alan Pevec	Alan Pevec	Red Hat OpenStack	Openstack-cinder	CLOSED	Unspecified	Unspecified	Unspecified	2.0 (Folsom)
883844	2012/12/5 0:00	2013-07-04T02:55:43Z	Alan Pevec	Alan Pevec	Red Hat OpenStack	Openstack-nova	CLOSED	Unspecified	Unspecified	Unspecified	2.0 (Folsom)
884449	2012/12/6 0:00	2012-12-11T13:08:35Z	Nir Magnezi	Derek Higgins	Red Hat OpenStack	Openstack-packstack	CLOSED	Unspecified	Linux	Medium	2.0 (Folsom)
884595	2012/12/6 0:00	2016-04-26T23:14:47Z	Alan Pevec	Gary Kotton	Red Hat OpenStack	Openstack-nova	CLOSED	Unspecified	Unspecified	High	2.0 (Folsom)
884714	2012/12/6 0:00	2012-12-11T13:08:37Z	Nir Magnezi	Derek Higgins	Red Hat OpenStack	Openstack-packstack	CLOSED	Unspecified	Linux	High	2.0 (Folsom)
884748	2012/12/6 0:00	2013-11-27T14:55:41Z	Nir Magnezi	Francesco Vollero	Red Hat OpenStack	Openstack-packstack	CLOSED	Unspecified	Linux	Medium	2.0 (Folsom)
884757	2012/12/6 0:00	2012-12-11T13:08:40Z	Nir Magnezi	Derek Higgins	Red Hat OpenStack	Openstack-packstack	CLOSED	Unspecified	Linux	High	2.0 (Folsom)

(continued)

8.4 Numerical Examples

Table 8.2 (continued)

Bug ID	Opened	Changed	Reporter	Product	Component	Status	Resolution	Hardware	OS	Severity	Version
885431	2012/12/9 0:00	2012-12-09T12:49:18Z	Nir Magnezi	Derek Higgins	Red Hat OpenStack	Openstack-packstack	CLOSED	Unspecified	Linux	Low	2.0 (Folsom)
885529	2012/12/10 0:00	2019-09-09T16:35:54Z	Derek Higgins	Lon Hohberger	Red Hat OpenStack	Openstack-selinux	CLOSED	Unspecified	Unspecified	High	2.0 (Folsom)
885530	2012/12/10 0:00	2016-04-26T13:25:51Z	Derek Higgins	Alan Pevec	Red Hat OpenStack	Openstack-swift	CLOSED	Unspecified	Unspecified	Unspecified	2.0 (Folsom)
885793	2012/12/10 0:00	2016-04-26T18:56:07Z	Nir Magnezi	Flavio Percoco	Red Hat OpenStack	Openstack-packstack	CLOSED	Unspecified	Linux	Urgent	2.0 (Folsom)
885888	2012/12/10 0:00	2013-02-14T18:23:13Z	Pete Zaitcev	Alan Pevec	Red Hat OpenStack	Openstack-swift	CLOSED	Unspecified	Unspecified	Unspecified	Unspecified
886177	2012/12/11 0:00	2013-01-29T11:09:52Z	Derek Higgins	Martin Magr	Red Hat OpenStack	Openstack-packstack	CLOSED	Unspecified	Unspecified	Medium	2.0 (Folsom)
886272	2012/12/11 0:00	2019-09-09T14:35:32Z	Graeme Gillies	Brent Eagles	Red Hat OpenStack	Openstack-neutron	CLOSED	Unspecified	Unspecified	Unspecified	2.1
886292	2012/12/11 0:00	2019-09-09T13:46:53Z	Graeme Gillies	Nikola Dipanov	Red Hat OpenStack	Openstack-nova	CLOSED	Unspecified	Unspecified	Low	2.1
886541	2012/12/12 0:00	2019-09-10T14:07:44Z	Derek Higgins	Martin Magr	Red Hat OpenStack	Openstack-packstack	CLOSED	Unspecified	Unspecified	High	2.0 (Folsom)
886592	2012/12/12 0:00	2022-07-09T07:08:38Z	Nir Magnezi	Martin Magr	Red Hat OpenStack	Openstack-packstack	CLOSED	Unspecified	Linux	Medium	2.0 (Folsom)
886603	2012/12/12 0:00	2022-07-09T06:23:15Z	Nir Magnezi	Martin Magr	Red Hat OpenStack	Openstack-packstack	CLOSED	Unspecified	Linux	High	2.0 (Folsom)
887209	2012/12/14 0:00	2016-04-18T07:02:27Z	Giulio Fidente	RHOS Maint	Red Hat OpenStack	Openstack-nova	CLOSED	Unspecified	Unspecified	Unspecified	2.0 (Folsom)
887217	2012/12/14 0:00	2016-01-04T14:43:36Z	Pdraig Brady	Alan Pevec	Red Hat OpenStack	Openstack-nova	CLOSED	Unspecified	Unspecified	Medium	2.0 (Folsom)
887299	2012/12/14 0:00	2016-04-26T22:37:42Z	Derek Higgins	Martin Magr	Red Hat OpenStack	Openstack-packstack	CLOSED	Unspecified	Unspecified	High	2.0 (Folsom)

(continued)

Table 8.2 (continued)

Bug ID	Opened	Changed	Reporter	Product	Component	Status	Resolution	Hardware	OS	Severity	Version
887300	2012/12/14 0:00	2013-10-01T14:37:52Z	Derek Higgins	RHOS Maint	Red Hat OpenStack	Openstack-packstack	CLOSED	Unspecified	Unspecified	High	2.0 (Folsom)
887303	2012/12/14 0:00	2022-07-09T06:04:52Z	Perry Myers	Alan Pevec	Red Hat OpenStack	Openstack-nova	CLOSED	Unspecified	Unspecified	High	2.0 (Folsom)
887334	2012/12/14 0:00	2019-09-09T17:03:19Z	Dan Yocum	Nikola Dipanov	Red Hat OpenStack	Openstack-nova	CLOSED	Unspecified	Unspecified	Medium	1.0 (Essex)
887804	2012/12/17 0:00	2016-04-26T16:16:00Z	Nikola Dipanov	Eric Harney	Red Hat OpenStack	Openstack-cinder	CLOSED	Unspecified	Unspecified	Medium	1.0 (Essex)
887811	2012/12/17 0:00	2016-04-27T01:23:26Z	Nikola Dipanov	John Bresnahan	Red Hat OpenStack	Openstack-glance	CLOSED	Unspecified	Unspecified	Medium	1.0 (Essex)
887815	2012/12/17 0:00	2022-07-09T06:21:45Z	Nikola Dipanov	Alan Pevec	Red Hat OpenStack	Openstack-keystone	CLOSED	Unspecified	Unspecified	Medium	1.0 (Essex)
887818	2012/12/17 0:00	2016-04-26T20:59:45Z	Nikola Dipanov	Pete Zaitcev	Red Hat OpenStack	Openstack-swift	CLOSED	Unspecified	Unspecified	Medium	1.0 (Essex)
887910	2012/12/17 0:00	2016-04-26T16:05:34Z	Perry Myers	RHOS Maint	Red Hat OpenStack	Openstack-nova	CLOSED	Unspecified	Unspecified	Medium	1.0 (Essex)
887956	2012/12/17 0:00	2016-04-26T19:36:23Z	Derek Higgins	Derek Higgins	Red Hat OpenStack	Openstack-packstack	CLOSED	Unspecified	Unspecified	Unspecified	2.0 (Folsom)
887968	2012/12/17 0:00	2022-07-09T07:18:05Z	Perry Myers	Daniel Berrang	Red Hat OpenStack	Openstack-nova	CLOSED	Unspecified	Unspecified	Medium	1.0 (Essex)
888066	2012/12/17 0:00	2015-06-04T21:50:11Z	Russell Bryant	Adam Young	Red Hat OpenStack	Openstack-keystone	CLOSED	Unspecified	Unspecified	Unspecified	2.1
888077	2012/12/17 0:00	2022-07-09T06:27:33Z	Russell Bryant	John Bresnahan	Red Hat OpenStack	Openstack-glance	CLOSED	Unspecified	Unspecified	Medium	2.1
888082	2012/12/18 0:00	2013-07-04T02:55:44Z	Russell Bryant	RHOS Maint	Red Hat OpenStack	Openstack-glance	CLOSED	Unspecified	Unspecified	Unspecified	2.1

(continued)

8.4 Numerical Examples

Table 8.2 (continued)

Bug ID	Opened	Changed	Reporter	Product	Component	Status	Resolution	Hardware	OS	Severity	Version
888091	2012/12/18 0:00	2016-04-26T18:58:35Z	John Bresnahan	John Bresnahan	Red Hat OpenStack	Openstack-glance	CLOSED	Unspecified	Unspecified	Low	2.0 (Folsom)
888098	2012/12/18 0:00	2013-07-04T02:55:44Z	Russell Bryant	RHOS Maint	Red Hat OpenStack	Openstack-utils	CLOSED	Unspecified	Unspecified	Unspecified	2.1
888113	2012/12/18 0:00	2013-09-30T02:04:07Z	John Bresnahan	John Bresnahan	Red Hat OpenStack	Openstack-glance	CLOSED	Unspecified	Unspecified	Low	2.0 (Folsom)
888241	2012/12/18 0:00	2016-04-26T15:01:31Z	Eoghan Glynn	Derek Higgins	Red Hat OpenStack	Openstack-packstack	CLOSED	All	Linux	High	2.1
888244	2012/12/18 0:00	2013-07-04T02:55:45Z	Martin Magr	Adam Young	Red Hat OpenStack	Openstack-keystone	CLOSED	x86_64	Linux	Low	2.0 (Folsom)
888281	2012/12/18 0:00	2016-04-26T22:13:04Z	Yaniv Kaul	Derek Higgins	Red Hat OpenStack	Openstack-packstack	CLOSED	Unspecified	Unspecified	High	2.0 (Folsom)
888298	2012/12/18 0:00	2016-04-26T19:49:58Z	Yaniv Kaul	RHOS Maint	Red Hat OpenStack	Openstack-packstack	CLOSED	Unspecified	Unspecified	Medium	2.0 (Folsom)
888303	2012/12/18 0:00	2019-09-10T14:10:46Z	Yaniv Kaul	Derek Higgins	Red Hat OpenStack	Openstack-packstack	CLOSED	Unspecified	Unspecified	Medium	2.0 (Folsom)
888328	2012/12/18 0:00	2022-07-09T06:09:44Z	Martin Magr	Alan Pevec	Red Hat OpenStack	Openstack-keystone	CLOSED	Unspecified	Unspecified	Unspecified	2.0 (Folsom)
888332	2012/12/18 0:00	2016-04-26T19:46:49Z	Yaniv Kaul	Derek Higgins	Red Hat OpenStack	Openstack-packstack	CLOSED	Unspecified	Unspecified	Medium	2.0 (Folsom)
888345	2012/12/18 0:00	2016-04-26T16:21:14Z	Yaniv Kaul	Derek Higgins	Red Hat OpenStack	Openstack-packstack	CLOSED	Unspecified	Unspecified	Urgent	2.0 (Folsom)
888357	2012/12/18 0:00	2016-04-26T21:12:57Z	Steven Hardy	Martin Magr	Red Hat OpenStack	Openstack-packstack	CLOSED	Unspecified	Unspecified	Medium	2.0 (Folsom)
888360	2012/12/18 0:00	2019-09-10T14:07:07Z	Steven Hardy	Flavio Percoco	Red Hat OpenStack	Openstack-packstack	CLOSED	Unspecified	Unspecified	Medium	2.0 (Folsom)

(continued)

Table 8.2 (continued)

Bug ID	Opened	Changed	Reporter	Product	Component	Status	Resolution	Hardware	OS	Severity	Version
888376	2012/12/18 0:00	2016-04-26T23:52:39Z	Steven Hardy	Martin Magr	Red Hat OpenStack	Openstack-packstack	CLOSED	Unspecified	Unspecified	Medium	2.0 (Folsom)
888408	2012/12/18 0:00	2013-07-04T02:56:38Z	Martin Magr	RHOS Maint	Red Hat OpenStack	Openstack-glance	CLOSED	Unspecified	Unspecified	Unspecified	2.0 (Folsom)
888444	2012/12/18 0:00	2016-04-26T19:23:44Z	Steven Hardy	Martin Magr	Red Hat OpenStack	Openstack-packstack	CLOSED	Unspecified	Unspecified	Unspecified	2.0 (Folsom)
888461	2012/12/18 0:00	2016-04-22T05:01:41Z	Russell Bryant	Nikola Dipanov	Red Hat OpenStack	Openstack-nova	CLOSED	Unspecified	Unspecified	Unspecified	2.1
888512	2012/12/18 0:00	2016-04-27T04:59:17Z	Yaniv Kaul	Derek Higgins	Red Hat OpenStack	Openstack-packstack	CLOSED	Unspecified	Unspecified	Medium	2.0 (Folsom)
888514	2012/12/18 0:00	2019-09-10T14:09:27Z	Yaniv Kaul	Derek Higgins	Red Hat OpenStack	Openstack-packstack	CLOSED	Unspecified	Unspecified	Low	2.0 (Folsom)
888537	2012/12/18 0:00	2013-09-30T02:04:07Z	John Bresnahan	RHOS Maint	Red Hat OpenStack	Openstack-glance	CLOSED	Unspecified	Unspecified	Unspecified	2.0 (Folsom)
888542	2012/12/18 0:00	2013-09-30T02:04:07Z	John Bresnahan	Eoghan Glynn	Red Hat OpenStack	Openstack-glance	CLOSED	Unspecified	Unspecified	Medium	2.0 (Folsom)
888545	2012/12/18 0:00	2013-09-30T02:04:07Z	John Bresnahan	Eoghan Glynn	Red Hat OpenStack	Openstack-glance	CLOSED	Unspecified	Unspecified	Medium	2.0 (Folsom)
888575	2012/12/18 0:00	2022-07-09T06:21:55Z	Russell Bryant	Adam Young	Red Hat OpenStack	Openstack-keystone	CLOSED	Unspecified	Unspecified	Low	2.1

(continued)

8.4 Numerical Examples

Table 8.2 (continued)

Bug ID	Opened	Changed	Reporter	Product	Component	Status	Resolution	Hardware	OS	Severity	Version
888683	2012/12/19 0:00	2016-04-26T15:00:53Z	Nir Magnezi	RHOS Maint	Red Hat OpenStack	Openstack-packstack	CLOSED	Unspecified	Linux	High	2.0 (Folsom)
888712	2012/12/19 0:00	2013-07-04T02:56:39Z	Eoghan Glynn	RHOS Maint	Red Hat OpenStack	Openstack-nova	CLOSED	Unspecified	Unspecified	Low	2.1
888725	2012/12/19 0:00	2022-07-09T06:23:50Z	Alan Pevec	Martin Magr	Red Hat OpenStack	Openstack-packstack	CLOSED	Unspecified	Unspecified	Medium	2.0 (Folsom)
888744	2012/12/19 0:00	2016-04-26T14:14:24Z	Matthias Runge	Derek Higgins	Red Hat OpenStack	Openstack-packstack	CLOSED	Unspecified	Unspecified	Low	2.1
888751	2012/12/19 0:00	2016-04-26T13:51:42Z	Matthias Runge	Martin Magr	Red Hat OpenStack	Openstack-packstack	CLOSED	Unspecified	Unspecified	Medium	2.1
888756	2012/12/19 0:00	2019-09-10T14:11:29Z	Yaniv Kaul	Derek Higgins	Red Hat OpenStack	Openstack-packstack	CLOSED	Unspecified	Unspecified	High	2.0 (Folsom)
888784	2012/12/19 0:00	2016-04-26T14:48:35Z	Ofer Blaut	RHOS Maint	Red Hat OpenStack	Openstack-packstack	CLOSED	Unspecified	Unspecified	Medium	2.0 (Folsom)
888931	2012/12/19 0:00	2019-09-10T14:09:48Z	Yaniv Kaul	Zane Bitter	Red Hat OpenStack	Openstack-nova	CLOSED	Unspecified	Unspecified	High	2.0 (Folsom)
888983	2012/12/19 0:00	2019-09-09T15:40:49Z	Graeme Gillies	RHOS Maint	Red Hat OpenStack	Openstack-nova	CLOSED	Unspecified	Unspecified	Low	2.1
889199	2012/12/20 0:00	2016-04-26T19:28:18Z	Nir Magnezi	Martin Magr	Red Hat OpenStack	Openstack-packstack	CLOSED	Unspecified	Linux	Medium	2.0 (Folsom)
889269	2012/12/20 0:00	2016-04-27T00:22:48Z	Nir Magnezi	Derek Higgins	Red Hat OpenStack	Openstack-packstack	CLOSED	Unspecified	Linux	Urgent	2.0 (Folsom)
889336	2012/12/20 0:00	2016-09-26T13:27:32Z	Dan Prince	John Bresnahan	Red Hat OpenStack	Openstack-glance	CLOSED	Unspecified	Unspecified	Medium	3
889756	2012/12/23 0:00	2012-12-30T13:07:50Z	Nir Magnezi	RHOS Maint	Red Hat OpenStack	Openstack-packstack	CLOSED	Unspecified	Linux	Medium	2.0 (Folsom)
889764	2012/12/23 0:00	2016-04-26T21:58:56Z	Gary Kotton	RHOS Maint	Red Hat OpenStack	Openstack-nova	CLOSED	Unspecified	Unspecified	Low	2.0 (Folsom)

(continued)

Table 8.2 (continued)

Bug ID	Opened	Changed	Reporter	Product	Component	Status	Resolution	Hardware	OS	Severity	Version
889766	2012/12/23 0:00	2016-04-27T02:00:07Z	Gary Kotton	Eoghan Glynn	Red Hat OpenStack	Openstack-glance	CLOSED	Unspecified	Unspecified	Unspecified	2.0 (Folsom)
889782	2012/12/23 0:00	2019-09-10T14:12:35Z	Gary Kotton	Lon Hohberger	Red Hat OpenStack	Openstack-selinux	CLOSED	Unspecified	Unspecified	High	2.0 (Folsom)
889868	2012/12/23 0:00	2022-07-09T06:08:11Z	Chris Wright	Gary Kotton	Red Hat OpenStack	Openstack-nova	CLOSED	Unspecified	Unspecified	High	2.0 (Folsom)
890058	2012/12/24 0:00	2016-04-26T18:45:14Z	Yaniv Kaul	Flavio Percoco	Red Hat OpenStack	Openstack-packstack	CLOSED	Unspecified	Unspecified	Low	2.0 (Folsom)
890168	2012/12/25 0:00	2013-01-03T16:35:58Z	Nir Magnezi	RHOS Maint	Red Hat OpenStack	Openstack-cinder	CLOSED	Unspecified	Linux	Medium	2.0 (Folsom)
890175	2012/12/25 0:00	2016-04-26T22:37:58Z	Nir Magnezi	Martin Magr	Red Hat OpenStack	Openstack-packstack	CLOSED	Unspecified	Linux	High	2.0 (Folsom)
890178	2012/12/25 0:00	2013-02-14T18:24:25Z	Perry Myers	RHOS Maint	Red Hat OpenStack	Openstack-cinder	CLOSED	Unspecified	Unspecified	Medium	1.0 (Essex)
890180	2012/12/25 0:00	2013-06-28T14:41:15Z	Perry Myers	RHOS Maint	Red Hat OpenStack	Openstack-glance	CLOSED	Unspecified	Unspecified	Medium	1.0 (Essex)
890181	2012/12/25 0:00	2014-09-08T05:39:06Z	Perry Myers	RHOS Maint	Red Hat OpenStack	Openstack-keystone	CLOSED	Unspecified	Unspecified	Medium	1.0 (Essex)
890182	2012/12/25 0:00	2014-09-08T05:39:06Z	Perry Myers	David Ripton	Red Hat OpenStack	Openstack-nova	CLOSED	Unspecified	Unspecified	Low	1.0 (Essex)
890183	2012/12/25 0:00	2016-04-18T06:44:55Z	Perry Myers	Martin Magr	Red Hat OpenStack	Openstack-packstack	CLOSED	Unspecified	Unspecified	Medium	1.0 (Essex)
890184	2012/12/25 0:00	2016-04-26T17:23:55Z	Perry Myers	Terry Wilson	Red Hat OpenStack	Openstack-neutron	CLOSED	Unspecified	Unspecified	Medium	1.0 (Essex)
890185	2012/12/25 0:00	2013-07-04T02:57:29Z	Perry Myers	RHOS Maint	Red Hat OpenStack	Openstack-swift	CLOSED	Unspecified	Unspecified	Medium	1.0 (Essex)
890196	2012/12/25 0:00	2016-04-26T17:23:08Z	Perry Myers	Steven Dake	Red Hat OpenStack	Openstack-heat	CLOSED	All	Linux	Medium	2.0 (Folsom)

(continued)

8.4 Numerical Examples

Table 8.2 (continued)

Bug ID	Opened	Changed	Reporter	Product	Component	Status	Resolution	Hardware	OS	Severity	Version
890197	2012/12/25 0:00	2016-04-19T01:27:50Z	Perry Myers	Steven Hardy	Red Hat OpenStack	Openstack-heat	CLOSED	All	Linux	Medium	2.0 (Folsom)
890198	2012/12/25 0:00	2015-06-04T21:50:12Z	Perry Myers	Angus Salkeld	Red Hat OpenStack	Openstack-ceilometer	CLOSED	All	Linux	Medium	2.0 (Folsom)
890200	2012/12/25 0:00	2016-04-26T13:47:48Z	Perry Myers	Eric Harney	Red Hat OpenStack	Openstack-cinder	CLOSED	All	Linux	Medium	2.0 (Folsom)
890202	2012/12/25 0:00	2014-08-26T18:29:44Z	Perry Myers	Eric Harney	Red Hat OpenStack	Openstack-cinder	CLOSED	All	Linux	Medium	2.0 (Folsom)
890206	2012/12/25 0:00	2013-07-04T02:57:29Z	Perry Myers	Flavio Percoco	Red Hat OpenStack	Openstack-glance	CLOSED	All	Linux	Medium	2.0 (Folsom)
890207	2012/12/25 0:00	2013-05-29T15:03:20Z	Perry Myers	Flavio Percoco	Red Hat OpenStack	Openstack-glance	CLOSED	All	Linux	Medium	2.0 (Folsom)
890208	2012/12/25 0:00	2014-10-07T00:16:10Z	Perry Myers	RHOS Maint	Red Hat OpenStack	Openstack-glance	CLOSED	All	Linux	Medium	4
890209	2012/12/25 0:00	2019-09-09T14:45:43Z	Perry Myers	John Bresnahan	Red Hat OpenStack	Openstack-glance	CLOSED	All	Linux	Medium	2.0 (Folsom)
890210	2012/12/25 0:00	2013-12-24T15:22:49Z	Perry Myers	RHOS Maint	Red Hat OpenStack	Openstack-glance	CLOSED	All	Linux	Medium	4
890211	2012/12/25 0:00	2016-04-26T13:58:39Z	Perry Myers	Flavio Percoco	Red Hat OpenStack	Openstack-glance	CLOSED	All	Linux	Medium	2.0 (Folsom)
890295	2012/12/26 0:00	2022-07-09T06:01:38Z	Yaniv Kaul	Martin Magr	Red Hat OpenStack	Openstack-packstack	CLOSED	Unspecified	Unspecified	Medium	2.0 (Folsom)
890330	2012/12/26 0:00	2013-07-04T02:57:29Z	Ofer Blaut	RHOS Maint	Red Hat OpenStack	Openstack-nova	CLOSED	Unspecified	Unspecified	Medium	2.0 (Folsom)
890354	2012/12/26 0:00	2013-07-04T02:57:29Z	Yaniv Kaul	Martin Magr	Red Hat OpenStack	Openstack-packstack	CLOSED	Unspecified	Unspecified	Unspecified	2.0 (Folsom)
890512	2012/12/27 0:00	2022-07-09T06:08:51Z	Ofer Blaut	Brent Eagles	Red Hat OpenStack	Openstack-nova	CLOSED	Unspecified	Unspecified	High	2.0 (Folsom)

(continued)

Table 8.2 (continued)

Bug ID	Opened	Changed	Reporter	Product	Component	Status	Resolution	Hardware	OS	Severity	Version
890521	2012/12/27 0:00	2013-01-10T21:20:40Z	Nir Magnezi	RHOS Maint	Red Hat OpenStack	Openstack-packstack	CLOSED	Unspecified	Linux	Medium	2.0 (Folsom)
890527	2012/12/27 0:00	2016-04-26T16:06:31Z	Ofer Blaut	RHOS Maint	Red Hat OpenStack	Openstack-neutron	CLOSED	Unspecified	Unspecified	Low	2.0 (Folsom)
890539	2012/12/27 0:00	2016-04-26T13:22:49Z	Yaniv Kaul	Eoghan Glynn	Red Hat OpenStack	Openstack-glance	CLOSED	Unspecified	Unspecified	Low	2.0 (Folsom)
890822	2012/12/30 0:00	2016-04-26T16:03:40Z	Nir Magnezi	Martin Magr	Red Hat OpenStack	Openstack-packstack	CLOSED	Unspecified	Linux	High	2.0 (Folsom)
890841	2012/12/30 0:00	2016-04-26T20:47:21Z	Nir Magnezi	Derek Higgins	Red Hat OpenStack	Openstack-packstack	CLOSED	Unspecified	Linux	High	2.0 (Folsom)
890968	2012/12/31 0:00	2019-09-09T13:04:30Z	Ofer Blaut	Gary Kotton	Red Hat OpenStack	Openstack-nova	CLOSED	Unspecified	Unspecified	Medium	2.0 (Folsom)
890978	2012/12/31 0:00	2019-09-09T15:37:24Z	Ofer Blaut	Gary Kotton	Red Hat OpenStack	Openstack-nova	CLOSED	Unspecified	Unspecified	Medium	2.0 (Folsom)
890994	2012/12/31 0:00	2016-04-27T01:38:24Z	Ofer Blaut	Nikola Dipanov	Red Hat OpenStack	Openstack-nova	CLOSED	Unspecified	Unspecified	High	2.0 (Folsom)

(continued)

8.4 Numerical Examples

Table 8.2 (continued)

Bug ID	Opened	Changed	Reporter	Product	Component	Status	Resolution	Hardware	OS	Severity	Version
891051	2013/1/1 0:00	2016-04-26T13:57:43Z	Nir Magnezi	Russell Bryant	Red Hat OpenStack	Openstack-nova	CLOSED	Unspecified	Linux	Low	2.0 (Folsom)
891347	2013/1/2 0:00	2022-09-09T06:10:05Z	Perry Myers	Pdraig Brady	Red Hat OpenStack	Openstack-nova	CLOSED	Unspecified	Unspecified	Unspecified	2.0 (Folsom)
891349	2013/1/2 0:00	2022-07-09T06:21:41Z	Mark McLoughlin	Nikola Dipanov	Red Hat OpenStack	Openstack-nova	CLOSED	Unspecified	Unspecified	Unspecified	2.0 (Folsom)
891350	2013/1/2 0:00	2016-01-04T14:42:03Z	Perry Myers	Pdraig Brady	Red Hat OpenStack	Openstack-nova	CLOSED	Unspecified	Unspecified	Unspecified	2.0 (Folsom)
891408	2013/1/2 0:00	2016-04-26T19:08:08Z	Perry Myers	Nikola Dipanov	Red Hat OpenStack	Openstack-nova	CLOSED	Unspecified	Unspecified	Medium	2.0 (Folsom)
891417	2013/1/2 0:00	2013-01-14T22:53:31Z	Perry Myers	Derek Higgins	Red Hat OpenStack	Openstack-packstack	CLOSED	Unspecified	Unspecified	High	2.0 (Folsom)
891420	2013/1/2 0:00	2022-07-09T06:09:14Z	Perry Myers	Pdraig Brady	Red Hat OpenStack	Openstack-nova	CLOSED	Unspecified	Unspecified	Low	2.0 (Folsom)
891422	2013/1/2 0:00	2019-09-09T17:03:02Z	Perry Myers	Nikola Dipanov	Red Hat OpenStack	Openstack-nova	CLOSED	Unspecified	Unspecified	Low	2.0 (Folsom)
891509	2013/1/3 0:00	2016-04-26T14:21:54Z	Attila Fazekas	Eric Harney	Red Hat OpenStack	Openstack-cinder	CLOSED	Unspecified	Unspecified	Medium	2.0 (Folsom)
891700	2013/1/3 0:00	2013-01-30T20:16:07Z	Dan Radez	Dan Radez	Red Hat OpenStack	Openstack-packstack	CLOSED	Unspecified	Unspecified	Unspecified	Unspecified
892242	2013/1/6 0:00	2013-01-06T05:16:02Z	Perry Myers	RHOS Maint	Red Hat OpenStack	Openstack-packstack	CLOSED	Unspecified	Unspecified	Low	1.0 (Essex)
892247	2013/1/6 0:00	2022-07-09T06:23:40Z	Perry Myers	Martin Magr	Red Hat OpenStack	Openstack-packstack	CLOSED	Unspecified	Unspecified	Low	2.0 (Folsom)
892249	2013/1/6 0:00	2013-01-10T21:07:33Z	Perry Myers	RHOS Maint	Red Hat OpenStack	Openstack-packstack	CLOSED	Unspecified	Unspecified	Medium	2.0 (Folsom)
892300	2013/1/6 0:00	2013-07-04T02:58:25Z	Yaniv Kaul	RHOS Maint	Red Hat OpenStack	Openstack-glance	CLOSED	Unspecified	Unspecified	Medium	2.0 (Folsom)

(continued)

Table 8.2 (continued)

Bug ID	Opened	Changed	Reporter	Product	Component	Status	Resolution	Hardware	OS	Severity	Version
892318	2013/1/6 0:00	2013-01-29T11:10:51Z	Perry Myers	Martin Magr	Red Hat OpenStack	Openstack-packstack	CLOSED	Unspecified	Unspecified	Medium	2.0 (Folsom)
892327	2013/1/6 0:00	2016-04-26T18:25:59Z	Perry Myers	Adam Young	Red Hat OpenStack	Openstack-keystone	CLOSED	Unspecified	Unspecified	Unspecified	2.0 (Folsom)
892602	2013/1/7 0:00	2019-11-18T12:50:21Z	Mark McLoughlin	Sven Anderson	Red Hat OpenStack	Openstack-nova	CLOSED	Unspecified	Unspecified	Medium	2.1
892686	2013/1/7 0:00	2022-07-09T06:08:12Z	Perry Myers	Eric Harney	Red Hat OpenStack	Openstack-cinder	CLOSED	Unspecified	Unspecified	Medium	2.0 (Folsom)
892942	2013/1/8 0:00	2022-07-09T07:09:06Z	Omri Hochman	Martin Magr	Red Hat OpenStack	Openstack-packstack	CLOSED	x86_64	Linux	High	2.0 (Folsom)
893007	2013/1/8 0:00	2019-09-10T14:12:18Z	Omri Hochman	Martin Magr	Red Hat OpenStack	Openstack-packstack	CLOSED	x86_64	Linux	Medium	2.0 (Folsom)
893045	2013/1/8 0:00	2016-04-26T18:33:12Z	Yaniv Kaul	Flavio Percoco	Red Hat OpenStack	Openstack-glance	CLOSED	Unspecified	Unspecified	Medium	2.0 (Folsom)
893072	2013/1/8 0:00	2016-04-26T14:08:04Z	Yaniv Kaul	Flavio Percoco	Red Hat OpenStack	Openstack-glance	CLOSED	Unspecified	Unspecified	Medium	2.0 (Folsom)
893100	2013/1/8 0:00	2019-09-09T15:17:41Z	Nir Magnezi	Xavier Queralt	Red Hat OpenStack	Openstack-nova	CLOSED	Unspecified	Linux	Medium	2.0 (Folsom)
893107	2013/1/8 0:00	2022-07-09T06:23:29Z	Omri Hochman	Martin Magr	Red Hat OpenStack	Openstack-packstack	CLOSED	x86_64	Linux	Medium	2.0 (Folsom)
893374	2013/1/9 0:00	2019-09-09T13:56:12Z	Nikola Dipanov	Nikola Dipanov	Red Hat OpenStack	Openstack-nova	CLOSED	Unspecified	Unspecified	Unspecified	3
893379	2013/1/9 0:00	2019-09-09T16:37:27Z	Nikola Dipanov	Nikola Dipanov	Red Hat OpenStack	Openstack-nova	CLOSED	Unspecified	Unspecified	Unspecified	Unspecified
893449	2013/1/9 0:00	2019-09-09T16:48:16Z	Rami Vaknin	Nikola Dipanov	Red Hat OpenStack	Openstack-nova	CLOSED	Unspecified	Unspecified	High	Unspecified
893669	2013/1/9 0:00	2013-02-14T18:24:46Z	Dan Radez	Flavio Percoco	Red Hat OpenStack	Openstack-packstack	CLOSED	Unspecified	Unspecified	High	Unspecified

(continued)

8.4 Numerical Examples

Table 8.2 (continued)

Bug ID	Opened	Changed	Reporter	Product	Component	Status	Resolution	Hardware	OS	Severity	Version
893934	2013/1/10 0:00	2013-02-14T18:24:49Z	Nir Magnezi	Flavio Percoco	Red Hat OpenStack	Openstack-packstack	CLOSED	Unspecified	Linux	Medium	2.0 (Folsom)
893938	2013/1/10 0:00	2019-09-09T16:12:33Z	Attila Fazekas	Jakub Ruzicka	Red Hat OpenStack	Openstack-nova	CLOSED	Unspecified	Unspecified	Medium	2.0 (Folsom)
894460	2013/1/11 0:00	2015-06-04T21:50:12Z	Perry Myers	Lukas Bezdicka	Red Hat OpenStack	Openstack-packstack	CLOSED	Unspecified	Unspecified	Low	2.0 (Folsom)
894712	2013/1/13 0:00	2013-02-07T23:53:54Z	Omri Hochman	RHOS Maint	Red Hat OpenStack	Openstack-packstack	CLOSED	x86_64	Linux	Urgent	2.0 (Folsom)
894715	2013/1/13 0:00	2019-09-09T15:47:30Z	Rami Vaknin	RHOS Maint	Red Hat OpenStack	Openstack-nova	CLOSED	Unspecified	Unspecified	Urgent	2.0 (Folsom)
894733	2013/1/13 0:00	2022-07-09T06:24:09Z	Yaniv Kaul	Martin Magr	Red Hat OpenStack	Openstack-packstack	CLOSED	Unspecified	Unspecified	Medium	2.0 (Folsom)
894789	2013/1/13 0:00	2019-06-13T07:52:44Z	Perry Myers	RHOS Maint	Red Hat OpenStack	Openstack-neutron	CLOSED	All	Linux	High	2.0 (Folsom)
894794	2013/1/13 0:00	2016-04-27T03:21:22Z	Perry Myers	RHOS Maint	Red Hat OpenStack	Openstack-keystone	CLOSED	All	Linux	Medium	2.0 (Folsom)
894795	2013/1/13 0:00	2016-04-26T17:07:41Z	Perry Myers	RHOS Maint	Red Hat OpenStack	Openstack-keystone	CLOSED	All	Linux	Medium	2.0 (Folsom)
894796	2013/1/13 0:00	2016-04-26T14:16:13Z	Perry Myers	RHOS Maint	Red Hat OpenStack	Openstack-keystone	CLOSED	All	Linux	Medium	2.0 (Folsom)
894812	2013/1/13 0:00	2019-09-09T17:04:47Z	Perry Myers	David Ripton	Red Hat OpenStack	Openstack-nova	CLOSED	All	Linux	Low	2.0 (Folsom)
894813	2013/1/13 0:00	2019-09-09T14:33:48Z	Perry Myers	Solly Ross	Red Hat OpenStack	Openstack-nova	CLOSED	All	Linux	Medium	2.0 (Folsom)
894814	2013/1/13 0:00	2019-09-09T15:27:38Z	Perry Myers	RHOS Maint	Red Hat OpenStack	Openstack-nova	CLOSED	All	Linux	Medium	2.0 (Folsom)
894815	2013/1/13 0:00	2016-04-26T17:56:33Z	Perry Myers	Steven Hardy	Red Hat OpenStack	Openstack-neutron	CLOSED	All	Linux	Medium	2.0 (Folsom)

(continued)

Table 8.2 (continued)

Bug ID	Opened	Changed	Reporter	Product	Component	Status	Resolution	Hardware	OS	Severity	Version
894820	2013/1/13 0:00	2019-09-09T16:27:33Z	Perry Myers	Nikola Dipanov	Red Hat OpenStack	Openstack-nova	CLOSED	All	Linux	Medium	2.0 (Folsom)
894824	2013/1/13 0:00	2019-09-09T14:00:46Z	Perry Myers	Dave Allan	Red Hat OpenStack	Openstack-nova	CLOSED	All	Linux	Medium	3
894826	2013/1/13 0:00	2013-01-30T16:09:07Z	Perry Myers	RHOS Maint	Red Hat OpenStack	Openstack-packstack	CLOSED	All	Linux	Medium	2.0 (Folsom)
894827	2013/1/13 0:00	2013-04-25T17:04:15Z	Perry Myers	RHOS Maint	Red Hat OpenStack	Openstack-packstack	CLOSED	All	Linux	Medium	2.0 (Folsom)
894828	2013/1/13 0:00	2013-06-11T18:50:07Z	Perry Myers	Terry Wilson	Red Hat OpenStack	Openstack-packstack	CLOSED	All	Linux	Medium	2.0 (Folsom)
894841	2013/1/13 0:00	2016-04-26T20:49:39Z	Perry Myers	lpeer	Red Hat OpenStack	Openstack-neutron	CLOSED	All	Linux	Medium	2.0 (Folsom)
894843	2013/1/13 0:00	2016-04-26T13:55:51Z	Perry Myers	RHOS Maint	Red Hat OpenStack	Openstack-neutron	CLOSED	All	Linux	Medium	2.0 (Folsom)
894845	2013/1/13 0:00	2019-09-09T14:52:34Z	Perry Myers	Gary Kotton	Red Hat OpenStack	Openstack-nova	CLOSED	All	Linux	Medium	2.0 (Folsom)

(continued)

8.4 Numerical Examples

Table 8.2 (continued)

Bug ID	Opened	Changed	Reporter	Product	Component	Status	Resolution	Hardware	OS	Severity	Version
894860	2013/1/13 0:00	2019-09-09T13:50:32Z	Perry Myers	Solly Ross	Red Hat OpenStack	Openstack-nova	CLOSED	All	Linux	Low	2.0 (Folsom)
894861	2013/1/13 0:00	2019-09-09T16:13:14Z	Perry Myers	RHOS Maint	Red Hat OpenStack	Openstack-nova	CLOSED	All	Linux	Medium	2.0 (Folsom)
894862	2013/1/13 0:00	2015-06-04T21:50:26Z	Perry Myers	Lon Hohberger	Red Hat OpenStack	Openstack-selinux	CLOSED	All	Linux	Medium	2.0 (Folsom)
894892	2013/1/13 0:00	2019-09-09T14:21:38Z	Perry Myers	Daniel Berrang	Red Hat OpenStack	Openstack-nova	CLOSED	All	Linux	Medium	2.0 (Folsom)
894917	2013/1/14 0:00	2019-09-09T15:36:10Z	Perry Myers	David Ripton	Red Hat OpenStack	Openstack-nova	CLOSED	All	Linux	Medium	2.0 (Folsom)
894924	2013/1/14 0:00	2016-04-27T05:26:16Z	Perry Myers	Bob Kukura	Red Hat OpenStack	Openstack-neutron	CLOSED	All	Linux	Medium	2.0 (Folsom)
894925	2013/1/14 0:00	2016-04-26T17:54:55Z	Perry Myers	Adam Young	Red Hat OpenStack	Openstack-keystone	CLOSED	All	Linux	Medium	2.0 (Folsom)
895042	2013/1/14 0:00	2016-04-27T01:22:22Z	Omri Hochman	Ivan Chavero	Red Hat OpenStack	Openstack-packstack	CLOSED	x86_64	Linux	Low	2.0 (Folsom)
895116	2013/1/14 0:00	2014-01-12T23:54:52Z	Rami Vaknin	Derek Higgins	Red Hat OpenStack	Openstack-packstack	CLOSED	Unspecified	Unspecified	Urgent	2.0 (Folsom)
895145	2013/1/14 0:00	2013-01-29T11:11:16Z	Stephen Gordon	Martin Magr	Red Hat OpenStack	Openstack-packstack	CLOSED	Unspecified	Unspecified	Unspecified	2.1
895251	2013/1/14 0:00	2016-04-26T14:50:09Z	Stephen Gordon	Martin Magr	Red Hat OpenStack	Openstack-packstack	CLOSED	Unspecified	Unspecified	Unspecified	2.1
895588	2013/1/15 0:00	2019-09-09T16:13:03Z	Stephen Gordon	Russell Bryant	Red Hat OpenStack	Openstack-nova	CLOSED	Unspecified	Unspecified	Unspecified	3
895804	2013/1/16 0:00	2016-04-26T14:31:46Z	Mike McCune	Martin Magr	Red Hat OpenStack	Openstack-packstack	CLOSED	Unspecified	Unspecified	Unspecified	Unspecified
896085	2013/1/16 0:00	2019-09-09T15:32:39Z	Rami Vaknin	David Ripton	Red Hat OpenStack	Openstack-nova	CLOSED	Unspecified	Unspecified	Low	2.0 (Folsom)

(continued)

Table 8.2 (continued)

Bug ID	Opened	Changed	Reporter	Product	Component	Status	Resolution	Hardware	OS	Severity	Version
896107	2013/1/16 0:00	2013-01-29T11:11:22Z	Nir Magnezi	Martin Magr	Red Hat OpenStack	Openstack-packstack	CLOSED	Unspecified	Linux	Urgent	2.0 (Folsom)
896115	2013/1/16 0:00	2019-09-09T16:26:40Z	John Bresnahan	John Bresnahan	Red Hat OpenStack	Openstack-nova	CLOSED	Unspecified	Unspecified	Unspecified	Unspecified
896156	2013/1/16 0:00	2013-01-29T11:11:24Z	Dan Radez	Martin Magr	Red Hat OpenStack	Openstack-packstack	CLOSED	Unspecified	Unspecified	Unspecified	Unspecified
896181	2013/1/16 0:00	2013-01-29T11:11:27Z	Stephen Gordon	Martin Magr	Red Hat OpenStack	Openstack-packstack	CLOSED	Unspecified	Unspecified	Unspecified	2.1
896188	2013/1/16 0:00	2013-01-29T11:11:30Z	Ofer Blaut	Martin Magr	Red Hat OpenStack	Openstack-packstack	CLOSED	Unspecified	Unspecified	Medium	2.0 (Folsom)
896236	2013/1/16 0:00	2013-04-30T12:18:55Z	Stephen Gordon	RHOS Maint	Red Hat OpenStack	Openstack-packstack	CLOSED	Unspecified	Unspecified	Low	2.1
896489	2013/1/17 0:00	2019-09-09T15:05:59Z	Rami Vaknin	RHOS Maint	Red Hat OpenStack	Openstack-nova	CLOSED	Unspecified	Unspecified	High	2.0 (Folsom)
896494	2013/1/17 0:00	2013-01-29T11:11:36Z	Nir Magnezi	Martin Magr	Red Hat OpenStack	Openstack-packstack	CLOSED	Unspecified	Linux	Medium	2.0 (Folsom)
896586	2013/1/17 0:00	2019-09-09T13:54:07Z	Rami Vaknin	Nikola Dipanov	Red Hat OpenStack	Openstack-nova	CLOSED	Unspecified	Unspecified	High	2.0 (Folsom)
896618	2013/1/17 0:00	2022-07-09T06:23:58Z	Stephen Gordon	Martin Magr	Red Hat OpenStack	Openstack-packstack	CLOSED	Unspecified	Unspecified	Unspecified	2.1
896619	2013/1/17 0:00	2019-09-09T14:11:45Z	Eric Harney	Nikola Dipanov	Red Hat OpenStack	Openstack-nova	CLOSED	Unspecified	Unspecified	Unspecified	2.1
896642	2013/1/17 0:00	2015-08-21T00:33:50Z	Nir Magnezi	Martin Magr	Red Hat OpenStack	Openstack-packstack	CLOSED	Unspecified	Linux	Urgent	2.0 (Folsom)
896706	2013/1/17 0:00	2019-09-09T15:44:25Z	Perry Myers	Lon Hohberger	Red Hat OpenStack	Openstack-selinux	CLOSED	x86_64	Linux	Urgent	2.0 (Folsom)
901345	2013/1/18 0:00	2013-01-29T11:11:42Z	Derek Higgins	Martin Magr	Red Hat OpenStack	Openstack-packstack	CLOSED	Unspecified	Unspecified	Medium	2.0 (Folsom)

(continued)

8.4 Numerical Examples

Table 8.2 (continued)

Bug ID	Opened	Changed	Reporter	Product	Component	Status	Resolution	Hardware	OS	Severity	Version
901955	2013/1/20 0:00	2016-04-27T01:18:09Z	Yaniv Kaul	RHOS Maint	Red Hat OpenStack	Openstack-keystone	CLOSED	Unspecified	Unspecified	Medium	2.0 (Folsom)
901968	2013/1/20 0:00	2019-09-09T16:09:15Z	Rami Vaknin	Brent Eagles	Red Hat OpenStack	Openstack-nova	CLOSED	Unspecified	Unspecified	High	2.0 (Folsom)
902020	2013/1/20 0:00	2019-09-10T14:11:40Z	Ofer Blaut	Flavio Percoco	Red Hat OpenStack	Openstack-packstack	CLOSED	Unspecified	Unspecified	Urgent	2.0 (Folsom)
902293	2013/1/21 0:00	2019-09-09T13:20:23Z	Julie Pichon	Eoghan Glynn	Red Hat OpenStack	Openstack-nova	CLOSED	Unspecified	Linux	Unspecified	2.1
902370	2013/1/21 0:00	2013-05-21T08:43:14Z	Nir Magnezi	RHOS Maint	Red Hat OpenStack	Openstack-packstack	CLOSED	Unspecified	Linux	High	2.1
902409	2013/1/21 0:00	2022-07-09T06:09:43Z	Rami Vaknin	Brent Eagles	Red Hat OpenStack	Openstack-nova	CLOSED	Unspecified	Unspecified	Medium	2.0 (Folsom)
902485	2013/1/21 0:00	2013-02-14T18:25:01Z	Eric Harney	Martin Magr	Red Hat OpenStack	Openstack-packstack	CLOSED	Unspecified	Unspecified	High	2.1
902546	2013/1/21 0:00	2016-04-26T21:58:02Z	Stephen Gordon	Alvaro Lopez Ortega	Red Hat OpenStack	Openstack-packstack	CLOSED	Unspecified	Unspecified	Low	3
902703	2013/1/22 0:00	2013-02-14T18:25:04Z	Martin Magr	Martin Magr	Red Hat OpenStack	Openstack-packstack	CLOSED	Unspecified	Unspecified	Low	2.0 (Folsom)
903095	2013/1/23 0:00	2013-05-13T10:37:17Z	Ofer Blaut	RHOS Maint	Red Hat OpenStack	Openstack-packstack	CLOSED	Unspecified	Unspecified	Low	2.0 (Folsom)
903124	2013/1/23 0:00	2019-09-09T13:53:16Z	Ofer Blaut	Nikola Dipanov	Red Hat OpenStack	Openstack-nova	CLOSED	Unspecified	Unspecified	Medium	2.0 (Folsom)
903153	2013/1/23 0:00	2013-02-14T18:25:07Z	Nir Magnezi	Flavio Percoco	Red Hat OpenStack	Openstack-packstack	CLOSED	Unspecified	Linux	Medium	2.1
903187	2013/1/23 0:00	2022-07-09T06:04:29Z	Flavio Percoco	Martin Magr	Red Hat OpenStack	Openstack-packstack	CLOSED	Unspecified	Unspecified	Low	Unspecified
903191	2013/1/23 0:00	2013-02-14T18:25:10Z	Flavio Percoco	Flavio Percoco	Red Hat OpenStack	Openstack-packstack	CLOSED	Unspecified	Unspecified	Medium	Unspecified

(continued)

Table 8.2 (continued)

Bug ID	Opened	Changed	Reporter	Product	Component	Status	Resolution	Hardware	OS	Severity	Version
903502	2013/1/24 0:00	2022-07-09T06:23:48Z	Mark McLoughlin	Martin Magr	Red Hat OpenStack	Openstack-packstack	CLOSED	Unspecified	Unspecified	High	2.1
903504	2013/1/24 0:00	2019-09-09T13:38:05Z	Ofer Blaut	David Ripton	Red Hat OpenStack	Openstack-nova	CLOSED	Unspecified	Unspecified	Low	2.0 (Folsom)
903545	2013/1/24 0:00	2022-07-09T07:20:39Z	Alan Pevec	Martin Magr	Red Hat OpenStack	Openstack-packstack	CLOSED	Unspecified	Unspecified	High	2.0 (Folsom)
903547	2013/1/24 0:00	2019-09-09T15:50:43Z	Ofer Blaut	David Ripton	Red Hat OpenStack	Openstack-nova	CLOSED	Unspecified	Unspecified	Low	2.0 (Folsom)
903568	2013/1/24 0:00	2019-09-09T13:28:39Z	Ofer Blaut	RHOS Maint	Red Hat OpenStack	Openstack-nova	CLOSED	Unspecified	Unspecified	Medium	2.0 (Folsom)
903671	2013/1/24 0:00	2019-09-09T16:49:19Z	Rami Vaknin	Brent Eagles	Red Hat OpenStack	Openstack-nova	CLOSED	Unspecified	Unspecified	Medium	2.0 (Folsom)
903696	2013/1/24 0:00	2019-09-09T13:25:29Z	Rami Vaknin	RHOS Maint	Red Hat OpenStack	Openstack-nova	CLOSED	Unspecified	Unspecified	Medium	2.0 (Folsom)
903705	2013/1/24 0:00	2019-09-09T13:59:50Z	Rami Vaknin	RHOS Maint	Red Hat OpenStack	Openstack-nova	CLOSED	Unspecified	Unspecified	Medium	2.0 (Folsom)

(continued)

8.4 Numerical Examples

Table 8.2 (continued)

Bug ID	Opened	Changed	Reporter	Product	Component	Status	Resolution	Hardware	OS	Severity	Version
894861	2013-01-13T20:26:04Z	2019-09-09T16:13:14Z	Perry Myers	RHOS Maint	Red Hat OpenStack	Openstack-nova	CLOSED	All	Linux	Medium	2.0 (Folsom)
894862	2013-01-13T20:26:44Z	2015-06-04T21:50:26Z	Perry Myers	Lon Hohberger	Red Hat OpenStack	Openstack-selinux	CLOSED	All	Linux	Medium	2.0 (Folsom)
894892	2013-01-13T23:03:20Z	2019-09-09T14:21:38Z	Perry Myers	Daniel Berrang	Red Hat OpenStack	Openstack-nova	CLOSED	All	Linux	Medium	2.0 (Folsom)
894917	2013-01-14T01:41:21Z	2019-09-09T15:36:10Z	Perry Myers	David Ripton	Red Hat OpenStack	Openstack-nova	CLOSED	All	Linux	Medium	2.0 (Folsom)
894924	2013-01-14T02:41:38Z	2016-04-27T05:26:16Z	Perry Myers	Bob Kukura	Red Hat OpenStack	Openstack-neutron	CLOSED	All	Linux	Medium	2.0 (Folsom)
894925	2013-01-14T02:43:36Z	2016-04-26T17:54:55Z	Perry Myers	Adam Young	Red Hat OpenStack	Openstack-keystone	CLOSED	All	Linux	Medium	2.0 (Folsom)
895042	2013-01-14T11:09:22Z	2016-04-27T01:22:22Z	Omri Hochman	Ivan Chavero	Red Hat OpenStack	Openstack-packstack	CLOSED	x86_64	Linux	Low	2.0 (Folsom)
895116	2013-01-14T15:51:35Z	2014-01-12T23:54:52Z	Rami Vaknin	Derek Higgins	Red Hat OpenStack	Openstack-packstack	CLOSED	Unspecified	Unspecified	Urgent	2.1
895145	2013-01-14T17:17:29Z	2013-01-29T11:11:16Z	Stephen Gordon	Martin Magr	Red Hat OpenStack	Openstack-packstack	CLOSED	Unspecified	Unspecified	Unspecified	2.1
895251	2013-01-14T22:33:59Z	2016-04-26T14:50:09Z	Stephen Gordon	Martin Magr	Red Hat OpenStack	Openstack-packstack	CLOSED	Unspecified	Unspecified	Unspecified	2.1
895588	2013-01-15T15:32:43Z	2019-09-09T16:13:03Z	Stephen Gordon	Russell Bryant	Red Hat OpenStack	Openstack-nova	CLOSED	Unspecified	Unspecified	Unspecified	3
895804	2013-01-16T02:52:23Z	2016-04-26T14:31:46Z	Mike McCune	Martin Magr	Red Hat OpenStack	Openstack-packstack	CLOSED	Unspecified	Unspecified	Unspecified	Unspecified
896085	2013-01-16T15:08:07Z	2019-09-09T15:32:39Z	Rami Vaknin	David Ripton	Red Hat OpenStack	Openstack-nova	CLOSED	Unspecified	Unspecified	Low	2.0 (Folsom)
896107	2013-01-16T16:11:51Z	2013-01-29T11:11:22Z	Nir Magnezi	Martin Magr	Red Hat OpenStack	Openstack-packstack	CLOSED	Unspecified	Linux	Urgent	2.0 (Folsom)

(continued)

Table 8.2 (continued)

Bug ID	Opened	Changed	Reporter	Product	Component	Status	Resolution	Hardware	OS	Severity	Version
896115	2013-01-16T16:30:46Z	2019-09-09T16:26:40Z	John Bresnahan	John Bresnahan	Red Hat OpenStack	Openstack-nova	CLOSED	Unspecified	Unspecified	Unspecified	Unspecified
896156	2013-01-16T18:33:52Z	2013-01-29T11:11:24Z	Dan Radez	Martin Magr	Red Hat OpenStack	Openstack-packstack	CLOSED	Unspecified	Unspecified	Unspecified	Unspecified
896181	2013-01-16T19:34:30Z	2013-01-29T11:11:27Z	Stephen Gordon	Martin Magr	Red Hat OpenStack	Openstack-packstack	CLOSED	Unspecified	Unspecified	Unspecified	2.1
896188	2013-01-16T20:00:05Z	2013-01-29T11:11:30Z	Ofer Blaut	Martin Magr	Red Hat OpenStack	Openstack-packstack	CLOSED	Unspecified	Unspecified	Medium	2.0 (Folsom)
896236	2013-01-16T21:43:34Z	2013-04-30T12:18:55Z	Stephen Gordon	RHOS Maint	Red Hat OpenStack	Openstack-packstack	CLOSED	Unspecified	Unspecified	Low	2.1
896489	2013-01-17T12:21:01Z	2019-09-09T15:05:59Z	Rami Vaknin	RHOS Maint	Red Hat OpenStack	Openstack-nova	CLOSED	Unspecified	Unspecified	High	2.0 (Folsom)
896494	2013-01-17T12:39:49Z	2013-01-29T11:11:36Z	Nir Magnezi	Martin Magr	Red Hat OpenStack	Openstack-packstack	CLOSED	Unspecified	Linux	Medium	2.0 (Folsom)
896586	2013-01-17T15:08:29Z	2019-09-09T13:54:07Z	Rami Vaknin	Nikola Dipanov	Red Hat OpenStack	Openstack-nova	CLOSED	Unspecified	Unspecified	High	2.0 (Folsom)
896618	2013-01-17T15:56:05Z	2022-07-09T06:23:58Z	Stephen Gordon	Martin Magr	Red Hat OpenStack	Openstack-packstack	CLOSED	Unspecified	Unspecified	Unspecified	2.1
896619	2013-01-17T16:03:24Z	2019-09-09T14:11:45Z	Eric Harney	Nikola Dipanov	Red Hat OpenStack	Openstack-nova	CLOSED	Unspecified	Unspecified	Unspecified	2.1
896642	2013-01-17T16:35:17Z	2015-08-21T00:33:50Z	Nir Magnezi	Martin Magr	Red Hat OpenStack	Openstack-packstack	CLOSED	Unspecified	Linux	Urgent	2.0 (Folsom)
896706	2013-01-17T19:50:05Z	2019-09-09T15:44:25Z	Perry Myers	Lon Hohberger	Red Hat OpenStack	Openstack-selinux	CLOSED	x86_64	Linux	Urgent	2.0 (Folsom)
901345	2013-01-18T02:30:32Z	2013-01-29T11:11:42Z	Derek Higgins	Martin Magr	Red Hat OpenStack	Openstack-packstack	CLOSED	Unspecified	Unspecified	Medium	2.0 (Folsom)
901955	2013-01-20T07:26:36Z	2016-04-27T01:18:09Z	Yaniv Kaul	RHOS Maint	Red Hat OpenStack	Openstack-keystone	CLOSED	Unspecified	Unspecified	Medium	2.0 (Folsom)

(continued)

8.4 Numerical Examples

Table 8.2 (continued)

Bug ID	Opened	Changed	Reporter	Product	Component	Status	Resolution	Hardware	OS	Severity	Version
901968	2013-01-20T09:49:54Z	2019-09-09T16:09:15Z	Rami Vaknin	Brent Eagles	Red Hat OpenStack	Openstack-nova	CLOSED	Unspecified	Unspecified	High	2.0 (Folsom)
902020	2013-01-20T13:36:52Z	2019-09-10T14:11:40Z	Ofer Blaut	Flavio Percoco	Red Hat OpenStack	Openstack-packstack	CLOSED	Unspecified	Unspecified	Urgent	2.0 (Folsom)
902293	2013-01-21T10:19:46Z	2019-09-09T13:20:23Z	Julie Pichon	Eoghan Glynn	Red Hat OpenStack	Openstack-nova	CLOSED	Unspecified	Linux	Unspecified	2.1
902370	2013-01-21T14:15:43Z	2013-05-21T08:43:14Z	Nir Magnezi	RHOS Maint	Red Hat OpenStack	Openstack-packstack	CLOSED	Unspecified	Linux	High	2.1
902409	2013-01-21T15:43:09Z	2022-07-09T06:09:43Z	Rami Vaknin	Brent Eagles	Red Hat OpenStack	Openstack-nova	CLOSED	Unspecified	Unspecified	Medium	2.0 (Folsom)
902485	2013-01-21T18:59:06Z	2013-02-14T18:25:01Z	Eric Harney	Martin Magr	Red Hat OpenStack	Openstack-packstack	CLOSED	Unspecified	Unspecified	High	2.1
902546	2013-01-21T22:17:04Z	2016-04-26T21:58:02Z	Stephen Gordon	Alvaro Lopez Ortega	Red Hat OpenStack	Openstack-packstack	CLOSED	Unspecified	Unspecified	Low	3
902703	2013-01-22T10:03:45Z	2013-02-14T18:25:04Z	Martin Magr	Martin Magr	Red Hat OpenStack	Openstack-packstack	CLOSED	Unspecified	Unspecified	Low	2.0 (Folsom)
903095	2013-01-23T07:22:35Z	2013-05-13T10:37:17Z	Ofer Blaut	RHOS Maint	Red Hat OpenStack	Openstack-packstack	CLOSED	Unspecified	Unspecified	Low	2.0 (Folsom)
903124	2013-01-23T09:22:44Z	2019-09-09T13:53:16Z	Ofer Blaut	Nikola Dipanov	Red Hat OpenStack	Openstack-nova	CLOSED	Unspecified	Unspecified	Medium	2.0 (Folsom)
903153	2013-01-23T10:27:38Z	2013-02-14T18:25:07Z	Nir Magnezi	Flavio Percoco	Red Hat OpenStack	Openstack-packstack	CLOSED	Unspecified	Linux	Medium	2.1
903187	2013-01-23T12:34:32Z	2022-07-09T06:04:29Z	Flavio Percoco	Martin Magr	Red Hat OpenStack	Openstack-packstack	CLOSED	Unspecified	Unspecified	Low	Unspecified
903191	2013-01-23T12:43:42Z	2013-02-14T18:25:10Z	Flavio Percoco	Flavio Percoco	Red Hat OpenStack	Openstack-packstack	CLOSED	Unspecified	Unspecified	Medium	Unspecified

(continued)

Table 8.2 (continued)

Bug ID	Opened	Changed	Reporter	Product	Component	Status	Resolution	Hardware	OS	Severity	Version
903502	2013-01-24T08:20:09Z	2022-07-09T06:23:48Z	Mark McLoughlin	Martin Magr	Red Hat OpenStack	Openstack-packstack	CLOSED	Unspecified	Unspecified	High	2.1
903504	2013-01-24T08:25:48Z	2019-09-09T13:38:05Z	Ofer Blaut	David Ripton	Red Hat OpenStack	Openstack-nova	CLOSED	Unspecified	Unspecified	Low	2.0 (Folsom)
903545	2013-01-24T09:29:19Z	2022-07-09T07:20:39Z	Alan Pevec	Martin Magr	Red Hat OpenStack	Openstack-packstack	CLOSED	Unspecified	Unspecified	High	2.0 (Folsom)
903547	2013-01-24T09:32:22Z	2019-09-09T15:50:43Z	Ofer Blaut	David Ripton	Red Hat OpenStack	Openstack-nova	CLOSED	Unspecified	Unspecified	Low	2.0 (Folsom)
903568	2013-01-24T10:12:15Z	2019-09-09T13:28:39Z	Ofer Blaut	RHOS Maint	Red Hat OpenStack	Openstack-nova	CLOSED	Unspecified	Unspecified	Medium	2.0 (Folsom)
903671	2013-01-24T14:46:19Z	2019-09-09T16:49:19Z	Rami Vaknin	Brent Eagles	Red Hat OpenStack	Openstack-nova	CLOSED	Unspecified	Unspecified	Medium	2.0 (Folsom)
903696	2013-01-24T15:32:53Z	2019-09-09T13:25:29Z	Rami Vaknin	RHOS Maint	Red Hat OpenStack	Openstack-nova	CLOSED	Unspecified	Unspecified	Medium	2.0 (Folsom)
903705	2013-01-24T15:50:00Z	2019-09-09T13:59:50Z	Rami Vaknin	RHOS Maint	Red Hat OpenStack	Openstack-nova	CLOSED	Unspecified	Unspecified	Medium	2.0 (Folsom)
903719	2013-01-24T16:20:40Z	2016-04-26T22:08:04Z	Eric Harney	Derek Higgins	Red Hat OpenStack	Openstack-packstack	CLOSED	Unspecified	Unspecified	Unspecified	2.1

(continued)

Table 8.2 (continued)

Bug ID	Opened	Changed	Reporter	Product	Component	Status	Resolution	Hardware	OS	Severity	Version
903813	2013-01-24T21:25:42Z	2022-07-09T06:24:07Z	Stephen Gordon	Martin Magr	Red Hat OpenStack	Openstack-packstack	CLOSED	Unspecified	Unspecified	Unspecified	2.1
904274	2013-01-25T22:38:37Z	2016-04-26T14:42:41Z	Bob Kukura	Ipeer	Red Hat OpenStack	Openstack-neutron	CLOSED	Unspecified	Unspecified	Low	2.0 (Folsom)
904669	2013-01-26T23:52:03Z	2022-07-09T06:02:27Z	Perry Myers	Martin Magr	Red Hat OpenStack	Openstack-packstack	CLOSED	Unspecified	Unspecified	High	2.0 (Folsom)
904671	2013-01-27T00:22:38Z	2013-02-14T18:25:18Z	Perry Myers	Martin Magr	Red Hat OpenStack	Openstack-packstack	CLOSED	Unspecified	Unspecified	Urgent	2.0 (Folsom)
904712	2013-01-27T08:52:07Z	2013-02-28T11:04:37Z	Ofer Blaut	RHOS Maint	Red Hat OpenStack	Openstack-packstack	CLOSED	Unspecified	Unspecified	Low	2.0 (Folsom)
904713	2013-01-27T08:59:51Z	2019-09-09T13:19:27Z	Ofer Blaut	RHOS Maint	Red Hat OpenStack	Openstack-nova	CLOSED	Unspecified	Unspecified	Unspecified	2.0 (Folsom)
904730	2013-01-27T10:45:42Z	2019-09-09T14:29:25Z	Ofer Blaut	RHOS Maint	Red Hat OpenStack	Openstack-nova	CLOSED	Unspecified	Unspecified	High	2.0 (Folsom)
904751	2013-01-27T11:35:03Z	2019-09-09T14:08:31Z	Ofer Blaut	Brent Eagles	Red Hat OpenStack	Openstack-nova	CLOSED	Unspecified	Unspecified	High	2.0 (Folsom)
904800	2013-01-27T15:38:28Z	2016-04-26T17:58:14Z	Yaniv Kaul	RHOS Maint	Red Hat OpenStack	Openstack-cinder	CLOSED	Unspecified	Unspecified	Medium	2.1
904920	2013-01-28T07:51:44Z	2019-09-09T13:45:30Z	Yaniv Kaul	Russell Bryant	Red Hat OpenStack	Openstack-nova	CLOSED	Unspecified	Unspecified	Low	2.1
905081	2013-01-28T14:31:56Z	2022-07-09T07:21:08Z	Pavel Sedlk	Martin Magr	Red Hat OpenStack	Openstack-packstack	CLOSED	Unspecified	Unspecified	Unspecified	2.0 (Folsom)
905113	2013-01-28T15:47:09Z	2022-07-09T06:09:41Z	Russell Bryant	Russell Bryant	Red Hat OpenStack	Openstack-nova	CLOSED	Unspecified	Unspecified	Unspecified	2.1
905114	2013-01-28T15:50:18Z	2019-09-09T15:35:27Z	Russell Bryant	Eric Harney	Red Hat OpenStack	Openstack-cinder	CLOSED	Unspecified	Unspecified	Medium	2.1
905368	2013-01-29T09:27:50Z	2022-07-09T06:25:11Z	Derek Higgins	Martin Magr	Red Hat OpenStack	Openstack-packstack	CLOSED	Unspecified	Unspecified	High	2.0 (Folsom)

(continued)

Table 8.2 (continued)

Bug ID	Opened	Changed	Reporter	Product	Component	Status	Resolution	Hardware	OS	Severity	Version
905437	2013-01-29T12:53:08Z	2019-09-09T14:28:39Z	Ofer Blaut	David Ripton	Red Hat OpenStack	Openstack-nova	CLOSED	Unspecified	Unspecified	Low	2.0 (Folsom)
905516	2013-01-29T14:54:13Z	2022-07-09T07:09:23Z	Omri Hochman	Martin Magr	Red Hat OpenStack	Openstack-packstack	CLOSED	x86_64	Linux	High	2.0 (Folsom)
905737	2013-01-30T03:49:07Z	2022-07-09T06:02:46Z	Perry Myers	Martin Magr	Red Hat OpenStack	Openstack-packstack	CLOSED	Unspecified	Unspecified	Medium	2.0 (Folsom)
905842	2013-01-30T09:57:10Z	2022-07-09T06:24:28Z	Rami Vaknin	Martin Magr	Red Hat OpenStack	Openstack-packstack	CLOSED	Unspecified	Unspecified	High	2.0 (Folsom)
905931	2013-01-30T13:58:30Z	2016-04-26T16:30:18Z	Ofer Blaut	Adam Young	Red Hat OpenStack	Openstack-keystone	CLOSED	Unspecified	Unspecified	Low	2.0 (Folsom)
906006	2013-01-30T15:44:38Z	2022-07-09T07:09:47Z	Stephen Gordon	Martin Magr	Red Hat OpenStack	Openstack-packstack	CLOSED	Unspecified	Unspecified	Low	2.1
906038	2013-01-30T16:42:44Z	2015-08-21T00:33:50Z	Nir Magnezi	Flavio Percoco	Red Hat OpenStack	Openstack-packstack	CLOSED	Unspecified	Linux	Medium	2.1
906298	2013-01-31T12:04:20Z	2019-09-10T14:12:17Z	Jaroslav Henner	Pdraig Brady	Red Hat OpenStack	Openstack-utils	CLOSED	x86_64	Linux	Low	2.0 (Folsom)
906410	2013-01-31T15:37:46Z	2022-07-09T06:05:12Z	Flavio Percoco	Martin Magr	Red Hat OpenStack	Openstack-packstack	CLOSED	Unspecified	Unspecified	Unspecified	Unspecified
906471	2013-01-31T17:04:20Z	2013-06-11T15:28:05Z	Derek Higgins	Derek Higgins	Red Hat OpenStack	Openstack-packstack	CLOSED	Unspecified	Unspecified	High	2.0 (Folsom)
906783	2013-02-01T13:34:30Z	2022-07-09T06:09:58Z	Gary Kotton	RHOS Maint	Red Hat OpenStack	Openstack-nova	CLOSED	Unspecified	Unspecified	Medium	2.0 (Folsom)
907084	2013-02-03T08:46:59Z	2016-04-26T19:59:52Z	Ofer Blaut	Adam Young	Red Hat OpenStack	Openstack-keystone	CLOSED	Unspecified	Unspecified	High	2.0 (Folsom)
907088	2013-02-03T09:38:43Z	2019-09-09T16:23:51Z	Rami Vaknin	RHOS Maint	Red Hat OpenStack	Openstack-nova	CLOSED	Unspecified	Unspecified	High	2.0 (Folsom)
907178	2013-02-03T15:46:15Z	2022-07-09T06:10:32Z	Rami Vaknin	Brent Eagles	Red Hat OpenStack	Openstack-nova	CLOSED	Unspecified	Unspecified	Low	2.0 (Folsom)

(continued)

8.4 Numerical Examples

Table 8.2 (continued)

Bug ID	Opened	Changed	Reporter	Product	Component	Status	Resolution	Hardware	OS	Severity	Version
907259	2013-02-04T00:36:34Z	2013-09-30T02:04:08Z	Graeme Gillies	John Bresnahan	Red Hat OpenStack	Openstack-glance	CLOSED	Unspecified	Unspecified	Unspecified	2.1
907367	2013-02-04T08:29:45Z	2016-04-27T03:53:29Z	Gary Kotton	RHOS Maint	Red Hat OpenStack	Openstack-packstack	CLOSED	Unspecified	Unspecified	Unspecified	2.0 (Folsom)
907529	2013-02-04T16:07:36Z	2013-12-19T23:57:34Z	Yaniv Kaul	Giulio Fidente	Red Hat OpenStack	Openstack-packstack	CLOSED	Unspecified	Unspecified	Medium	2.1
907624	2013-02-04T20:43:13Z	2022-07-09T06:03:55Z	Stephen Gordon	Martin Magr	Red Hat OpenStack	Openstack-packstack	CLOSED	Unspecified	Unspecified	Low	2.1
907737	2013-02-05T07:15:12Z	2022-07-09T06:03:10Z	Gary Kotton	Martin Magr	Red Hat OpenStack	Openstack-packstack	CLOSED	Unspecified	Unspecified	Low	2.0 (Folsom)
908039	2013-02-05T17:25:58Z	2019-09-09T16:01:36Z	Yaniv Kaul	Brent Eagles	Red Hat OpenStack	Openstack-nova	CLOSED	Unspecified	Unspecified	Medium	2.1
908355	2013-02-06T13:51:14Z	2016-04-26T17:16:47Z	Pavel Sedlk	Adam Young	Red Hat OpenStack	Openstack-keystone	CLOSED	Unspecified	Unspecified	High	2.0 (Folsom)
908370	2013-02-06T14:33:11Z	2019-09-09T15:40:59Z	Ofer Blaut	David Ripton	Red Hat OpenStack	Openstack-nova	CLOSED	Unspecified	Unspecified	Low	2.0 (Folsom)
908373	2013-02-06T14:42:03Z	2022-07-09T06:10:24Z	Ofer Blaut	Gary Kotton	Red Hat OpenStack	Openstack-nova	CLOSED	Unspecified	Unspecified	Medium	2.0 (Folsom)
908379	2013-02-06T14:53:56Z	2016-04-26T21:05:18Z	Ofer Blaut	RHOS Maint	Red Hat OpenStack	Openstack-packstack	CLOSED	Unspecified	Unspecified	Low	2.0 (Folsom)
908514	2013-02-06T21:54:05Z	2019-09-09T15:37:26Z	james labocki	Brent Eagles	Red Hat OpenStack	Openstack-packstack	CLOSED	Unspecified	Unspecified	Medium	2.0 (Folsom)
908695	2013-02-07T10:55:11Z	2022-07-09T07:21:33Z	Omri Hochman	Martin Magr	Red Hat OpenStack	Openstack-packstack	CLOSED	x86_64	Linux	High	2.0 (Folsom)
908771	2013-02-07T13:38:28Z	2022-07-09T06:24:52Z	Gary Kotton	Martin Magr	Red Hat OpenStack	Openstack-packstack	CLOSED	Unspecified	Unspecified	Low	2.0 (Folsom)
908837	2013-02-07T16:18:19Z	2022-07-09T06:25:12Z	Jaroslav Henner	Martin Magr	Red Hat OpenStack	Openstack-packstack	CLOSED	x86_64	Linux	Medium	2.0 (Folsom)

(continued)

Table 8.2 (continued)

Bug ID	Opened	Changed	Reporter	Product	Component	Status	Resolution	Hardware	OS	Severity	Version
908838	2013-02-07T16:20:43Z	2022-07-09T06:25:30Z	Jaroslav Henner	Martin Magr	Red Hat OpenStack	Openstack-packstack	CLOSED	x86_64	Linux	Medium	2.0 (Folsom)
908846	2013-02-07T16:38:14Z	2022-07-09T06:25:21Z	Jaroslav Henner	Martin Magr	Red Hat OpenStack	Openstack-packstack	CLOSED	x86_64	Linux	Medium	2.0 (Folsom)
908892	2013-02-07T19:15:26Z	2016-04-26T21:08:28Z	Brent Eagles	RHOS Maint	Red Hat OpenStack	Openstack-packstack	CLOSED	Unspecified	Linux	Low	2.0 (Folsom)
908900	2013-02-07T20:08:56Z	2022-07-09T06:25:50Z	Jaroslav Henner	Martin Magr	Red Hat OpenStack	Openstack-packstack	CLOSED	x86_64	Linux	High	2.0 (Folsom)
908995	2013-02-08T01:37:26Z	2022-07-09T06:06:48Z	Alan Pevec	Alan Pevec	Red Hat OpenStack	Openstack-keystone	CLOSED	Unspecified	Unspecified	Unspecified	2.0 (Folsom)
909137	2013-02-08T11:50:58Z	2019-09-09T13:21:00Z	Pdraig Brady	Pdraig Brady	Red Hat OpenStack	Openstack-nova	CLOSED	Unspecified	Unspecified	Medium	2.0 (Folsom)
909286	2013-02-08T14:48:32Z	2019-09-09T14:08:15Z	Pdraig Brady	Pdraig Brady	Red Hat OpenStack	Openstack-nova	CLOSED	Unspecified	Unspecified	High	2.1
909418	2013-02-08T17:19:17Z	2016-01-04T14:46:18Z	Pdraig Brady	Eric Harney	Red Hat OpenStack	Openstack-cinder	CLOSED	Unspecified	Unspecified	High	2.1

(continued)

8.4 Numerical Examples

Table 8.2 (continued)

Bug ID	Opened	Changed	Reporter	Product	Component	Status	Resolution	Hardware	OS	Severity	Version
909418	2013-02-08T17:19:17Z	2016-01-04T14:46:18Z	draig Brady	Eric Harney	Red Hat OpenStack	Openstack-cinder	CLOSED	Unspecified	Unspecified	High	2.1
909424	2013-02-08T17:53:28Z	2016-04-26T14:03:08Z	Brent Eagles	Derek Higgins	Red Hat OpenStack	Openstack-packstack	CLOSED	Unspecified	Linux	Low	2.0 (Folsom)
909434	2013-02-08T18:37:59Z	2016-04-27T02:35:20Z	John Bresnahan	John Bresnahan	Red Hat OpenStack	Openstack-glance	CLOSED	Unspecified	Unspecified	Medium	2.0 (Folsom)
909438	2013-02-08T19:25:39Z	2013-02-26T12:47:38Z	Jaroslav Henner	Derek Higgins	Red Hat OpenStack	Openstack-packstack	CLOSED	x86_64	Linux	High	2.0 (Folsom)
909441	2013-02-08T19:34:01Z	2013-02-14T21:17:02Z	Jon Thomas	Martin Magr	Red Hat OpenStack	Openstack-packstack	CLOSED	Unspecified	Unspecified	Medium	2.0 (Folsom)
909460	2013-02-08T20:33:34Z	2016-04-26T18:17:10Z	Eric Harney	Eric Harney	Red Hat OpenStack	Openstack-cinder	CLOSED	Unspecified	Unspecified	Unspecified	2.1
909505	2013-02-09T00:36:34Z	2016-04-26T22:33:54Z	Derek Higgins	Martin Magr	Red Hat OpenStack	Openstack-packstack	CLOSED	Unspecified	Unspecified	Medium	2.0 (Folsom)
909653	2013-02-10T07:51:30Z	2019-09-09T14:19:22Z	Gary Kotton	Russell Bryant	Red Hat OpenStack	Openstack-nova	CLOSED	Unspecified	Unspecified	High	2.0 (Folsom)
909809	2013-02-11T07:19:52Z	2019-09-09T17:13:03Z	Ofer Blaut	Brent Eagles	Red Hat OpenStack	Openstack-nova	CLOSED	Unspecified	Unspecified	Medium	2.0 (Folsom)
909832	2013-02-11T09:27:54Z	2019-09-09T17:11:18Z	Gary Kotton	Brent Eagles	Red Hat OpenStack	Openstack-nova	CLOSED	Unspecified	Unspecified	Unspecified	2.0 (Folsom)
910089	2013-02-11T18:58:08Z	2022-07-09T07:22:23Z	Jaroslav Henner	Martin Magr	Red Hat OpenStack	Openstack-packstack	CLOSED	x86_64	Linux	High	2.0 (Folsom)
910210	2013-02-12T01:38:31Z	2022-07-09T06:25:33Z	Derek Higgins	Martin Magr	Red Hat OpenStack	Openstack-packstack	CLOSED	Unspecified	Unspecified	High	2.0 (Folsom)
910211	2013-02-12T01:43:07Z	2022-07-09T06:05:30Z	Derek Higgins	Martin Magr	Red Hat OpenStack	Openstack-packstack	CLOSED	Unspecified	Unspecified	High	2.0 (Folsom)

(continued)

Table 8.2 (continued)

Bug ID	Opened	Changed	Reporter	Product	Component	Status	Resolution	Hardware	OS	Severity	Version
910556	2013-02-12T20:45:46Z	2013-04-04T17:59:47Z	Jaroslav Henner	Derek Higgins	Red Hat OpenStack	Openstack-packstack	CLOSED	x86_64	Linux	High	2.0 (Folsom)
910619	2013-02-13T01:21:10Z	2019-09-09T16:15:20Z	Derek Higgins	RHOS Maint	Red Hat OpenStack	Openstack-nova	CLOSED	Unspecified	Unspecified	High	2.0 (Folsom)
910727	2013-02-13T11:47:47Z	2022-07-09T07:18:33Z	Jaroslav Henner	Russell Bryant	Red Hat OpenStack	Openstack-nova	CLOSED	x86_64	Linux	High	2.0 (Folsom)
910747	2013-02-13T13:22:51Z	2019-09-09T13:05:54Z	Omri Hochman	Xavier Queralt	Red Hat OpenStack	Openstack-nova	CLOSED	x86_64	Linux	Medium	2.0 (Folsom)
910806	2013-02-13T15:55:34Z	2019-09-09T15:08:18Z	Lon Hohberger	Lon Hohberger	Red Hat OpenStack	Openstack-nova	CLOSED	x86_64	Linux	Medium	2.0 (Folsom)
910818	2013-02-13T16:14:21Z	2022-07-09T06:03:33Z	Derek Higgins	Martin Magr	Red Hat OpenStack	Openstack-packstack	CLOSED	Unspecified	Unspecified	Unspecified	2.0 (Folsom)
911005	2013-02-14T08:08:13Z	2019-09-09T15:03:32Z	Ofer Blaut	David Ripton	Red Hat OpenStack	Openstack-nova	CLOSED	Unspecified	Unspecified	Urgent	2.0 (Folsom)
911103	2013-02-14T11:22:39Z	2022-07-09T06:10:56Z	draig Brady	draig Brady	Red Hat OpenStack	Openstack-nova	CLOSED	Unspecified	Unspecified	Unspecified	2.0 (Folsom)
911149	2013-02-14T13:43:53Z	2019-09-09T16:15:57Z	Ofer Blaut	Gary Kotton	Red Hat OpenStack	Openstack-nova	CLOSED	Unspecified	Unspecified	High	2.0 (Folsom)
911190	2013-02-14T14:43:07Z	2019-09-09T14:10:40Z	Omri Hochman	Nikola Dipanov	Red Hat OpenStack	Openstack-nova	CLOSED	x86_64	Linux	Medium	2.0 (Folsom)
911225	2013-02-14T15:35:02Z	2019-09-09T17:16:04Z	Omri Hochman	Russell Bryant	Red Hat OpenStack	Openstack-nova	CLOSED	x86_64	Linux	High	2.0 (Folsom)
911568	2013-02-15T11:34:27Z	2023-02-22T23:02:48Z	Flavio Percoco	Flavio Percoco	Red Hat OpenStack	Openstack-glance	CLOSED	Unspecified	Unspecified	Medium	Unspecified
911626	2013-02-15T13:55:33Z	2022-07-09T07:22:45Z	Jaroslav Henner	Martin Magr	Red Hat OpenStack	Openstack-packstack	CLOSED	x86_64	Linux	Low	2.0 (Folsom)

(continued)

8.4 Numerical Examples 115

Table 8.2 (continued)

Bug ID	Opened	Changed	Reporter	Product	Component	Status	Resolution	Hardware	OS	Severity	Version
911653	2013-02-15T14:37:47Z	2022-07-09T06:04:22Z	Jaroslav Henner	Martin Magr	Red Hat OpenStack	Openstack-packstack	CLOSED	x86_64	Linux	Urgent	2.0 (Folsom)
912006	2013-02-17T07:52:05Z	2022-07-09T06:25:52Z	Nir Magnezi	Martin Magr	Red Hat OpenStack	Openstack-packstack	CLOSED	Unspecified	Linux	High	2.1
912078	2013-02-17T16:45:54Z	2019-09-09T15:18:40Z	draig Brady	draig Brady	Red Hat OpenStack	Openstack-nova	CLOSED	Unspecified	Unspecified	Medium	2.0 (Folsom)
912143	2013-02-17T20:53:31Z	2016-04-27T05:45:50Z	Justin Clift	Eoghan Glynn	Red Hat OpenStack	Openstack-glance	CLOSED	Unspecified	Unspecified	Unspecified	2.0 (Folsom)
912284	2013-02-18T09:50:49Z	2022-07-09T06:22:44Z	Gary Kotton	Brent Eagles	Red Hat OpenStack	Openstack-nova	CLOSED	Unspecified	Unspecified	High	2.0 (Folsom)
912316	2013-02-18T11:56:16Z	2019-09-09T16:13:39Z	Omri Hochman	Brent Eagles	Red Hat OpenStack	Openstack-nova	CLOSED	x86_64	Linux	High	2.0 (Folsom)
912363	2013-02-18T14:19:27Z	2019-09-09T15:07:12Z	Omri Hochman	RHOS Maint	Red Hat OpenStack	Openstack-nova	CLOSED	x86_64	Linux	Medium	2.0 (Folsom)
912384	2013-02-18T15:04:54Z	2022-07-09T06:10:45Z	Eric Harney	Eric Harney	Red Hat OpenStack	Openstack-nova	CLOSED	Unspecified	Unspecified	Unspecified	2.1
912509	2013-02-18T20:45:07Z	2016-04-26T17:26:59Z	Russell Bryant	Eric Harney	Red Hat OpenStack	Openstack-cinder	CLOSED	Unspecified	Unspecified	Medium	2.1
912682	2013-02-19T12:15:25Z	2016-01-04T14:43:53Z	Tom Degroote	draig Brady	Red Hat OpenStack	Openstack-utils	CLOSED	x86_64	Linux	High	2.0 (Folsom)
912702	2013-02-19T13:12:17Z	2022-07-09T06:25:58Z	Perry Myers	Martin Magr	Red Hat OpenStack	Openstack-packstack	CLOSED	Unspecified	Unspecified	Low	2.0 (Folsom)
912744	2013-02-19T15:04:15Z	2019-09-09T13:44:30Z	Jaroslav Henner	Daniel Berrang 7?	Red Hat OpenStack	Openstack-nova	CLOSED	x86_64	Linux	Medium	2.0 (Folsom)
912745	2013-02-19T15:09:38Z	2022-07-09T06:26:26Z	Stephen Gordon	Martin Magr	Red Hat OpenStack	Openstack-packstack	CLOSED	Unspecified	Unspecified	Medium	2.1

(continued)

Table 8.2 (continued)

Bug ID	Opened	Changed	Reporter	Product	Component	Status	Resolution	Hardware	OS	Severity	Version
912768	2013-02-19T15:49:55Z	2022-07-09T06:26:21Z	Derek Higgins	Martin Magr	Red Hat OpenStack	Openstack-packstack	CLOSED	Unspecified	Unspecified	Medium	2.1
913195	2013-02-20T16:02:09Z	2016-04-26T14:12:43Z	Jaroslav Henner	Pete Zaitcev	Red Hat OpenStack	Openstack-swift	CLOSED	x86_64	Linux	Medium	2.0 (Folsom)
913197	2013-02-20T16:06:31Z	2019-09-09T14:09:58Z	Lon Hohberger	Lon Hohberger	Red Hat OpenStack	Openstack-selinux	CLOSED	x86_64	Linux	Urgent	2.1
913293	2013-02-20T21:20:33Z	2013-03-06T13:41:29Z	Jaroslav Henner	RHOS Maint	Red Hat OpenStack	Openstack-packstack	CLOSED	x86_64	Linux	Medium	2.0 (Folsom)
913513	2013-02-21T12:01:30Z	2016-01-04T14:46:42Z	Attila Fazekas	draig Brady	Red Hat OpenStack	Openstack-utils	CLOSED	Unspecified	Unspecified	Low	2.0 (Folsom)
913561	2013-02-21T13:34:33Z	2019-09-09T17:03:26Z	Attila Fazekas	Brent Eagles	Red Hat OpenStack	Openstack-nova	CLOSED	Unspecified	Unspecified	Low	2.0 (Folsom)
913613	2013-02-21T15:51:50Z	2022-07-09T06:09:56Z	draig Brady	draig Brady	Red Hat OpenStack	Openstack-nova	CLOSED	Unspecified	Unspecified	Medium	2.0 (Folsom)
913616	2013-02-21T15:55:43Z	2019-09-09T14:32:37Z	Daniel Berrang ??	Solly Ross	Red Hat OpenStack	Openstack-nova	CLOSED	Unspecified	Unspecified	Medium	3
913619	2013-02-21T15:58:21Z	2023-03-21T17:54:36Z	Daniel Berrang ??	OSP DFG:Compute	Red Hat OpenStack	Openstack-nova	CLOSED	x86_64	Linux	Low	3
914648	2013-02-22T12:01:09Z	2016-04-26T16:48:35Z	Jaroslav Henner	Martin Magr	Red Hat OpenStack	Openstack-packstack	CLOSED	x86_64	Linux	Medium	2.0 (Folsom)
914736	2013-02-22T16:30:05Z	2013-03-06T13:58:00Z	Dan Prince	RHOS Maint	Red Hat OpenStack	Openstack-packstack	CLOSED	Unspecified	Unspecified	High	2.1

(continued)

8.4 Numerical Examples

Table 8.2 (continued)

Bug ID	Opened	Changed	Reporter	Product	Component	Status	Resolution	Hardware	OS	Severity	Version
914759	2013-02-22T17:08:32Z	2022-07-09T06:11:11Z	Russell Bryant	Brent Eagles	Red Hat OpenStack	Openstack-nova	CLOSED	Unspecified	Unspecified	High	2.1
915274	2013-02-25T11:14:55Z	2022-07-09T06:22:38Z	Kashyap Chamarthy	Nikola Dipanov	Red Hat OpenStack	Openstack-nova	CLOSED	Unspecified	Unspecified	Unspecified	2.1
915365	2013-02-25T15:24:42Z	2013-12-19T23:57:58Z	Perry Myers	Martin Magr	Red Hat OpenStack	Openstack-packstack	CLOSED	Unspecified	Unspecified	Medium	2.0 (Folsom)
915382	2013-02-25T15:55:05Z	2022-07-09T06:26:17Z	Stephen Gordon	Martin Magr	Red Hat OpenStack	Openstack-packstack	CLOSED	Unspecified	Unspecified	Low	2.1
915383	2013-02-25T15:57:31Z	2013-04-04T18:00:08Z	Stephen Gordon	Martin Magr	Red Hat OpenStack	Openstack-packstack	CLOSED	Unspecified	Unspecified	Low	2.1
915445	2013-02-25T18:42:12Z	2022-07-09T06:05:38Z	Flavio Percoco	Flavio Percoco	Red Hat OpenStack	Openstack-glance	CLOSED	Unspecified	Unspecified	Unspecified	Unspecified
915881	2013-02-26T17:15:08Z	2016-04-26T16:31:55Z	Russell Bryant	Martin Magr	Red Hat OpenStack	Openstack-packstack	CLOSED	Unspecified	Unspecified	Unspecified	2.0 (Folsom)
915906	2013-02-26T18:35:28Z	2016-04-26T16:33:53Z	Ofer Blaut	Lon Hohberger	Red Hat OpenStack	Openstack-selinux	CLOSED	Unspecified	Unspecified	High	2.0 (Folsom)
915968	2013-02-26T21:29:12Z	2013-09-30T02:04:08Z	John Bresnahan	RHOS Maint	Red Hat OpenStack	Openstack-glance	CLOSED	Unspecified	Unspecified	Unspecified	2.0 (Folsom)
916174	2013-02-27T13:10:36Z	2022-07-09T06:03:49Z	Nikola Dipanov	Brent Eagles	Red Hat OpenStack	Openstack-nova	CLOSED	Unspecified	Unspecified	Unspecified	2.1
916176	2013-02-27T13:15:27Z	2022-07-09T06:03:49Z	Daniel Berrang	Daniel Berrang	Red Hat OpenStack	Openstack-nova	CLOSED	Unspecified	Unspecified	High	2.1
916241	2013-02-27T15:50:16Z	2022-07-09T06:10:15Z	Dan Prince	Nikola Dipanov	Red Hat OpenStack	Openstack-nova	CLOSED	Unspecified	Unspecified	Urgent	2.1
916284	2013-02-27T18:03:01Z	2016-04-26T19:40:49Z	Giulio Fidente	Eric Harney	Red Hat OpenStack	Openstack-cinder	CLOSED	Unspecified	Unspecified	High	2.1

(continued)

Table 8.2 (continued)

Bug ID	Opened	Changed	Reporter	Product	Component	Status	Resolution	Hardware	OS	Severity	Version
916301	2013-02-27T19:11:42Z	2016-04-26T16:31:37Z	Steve Reichard	Giulio Fidente	Red Hat OpenStack	Openstack-packstack	CLOSED	Unspecified	Unspecified	Medium	2.1
916307	2013-02-27T19:35:16Z	2013-02-28T08:00:42Z	Jon Thomas	RHOS Maint	Red Hat OpenStack	Openstack-packstack	CLOSED	Unspecified	Unspecified	Medium	2.0 (Folsom)
916329	2013-02-27T20:30:10Z	2016-04-26T13:41:22Z	Stephen Gordon	Steven Dake	Red Hat OpenStack	Openstack-utils	CLOSED	Unspecified	Unspecified	Low	2.1
916558	2013-02-28T10:15:36Z	2016-01-04T14:47:43Z	Attila Fazekas	draig Brady	Red Hat OpenStack	Openstack-utils	CLOSED	Unspecified	Unspecified	Low	2.0 (Folsom)
916567	2013-02-28T10:48:55Z	2016-01-04T14:47:03Z	Attila Fazekas	draig Brady	Red Hat OpenStack	Openstack-utils	CLOSED	Unspecified	Unspecified	Medium	2.0 (Folsom)
916605	2013-02-28T13:13:44Z	2013-04-04T18:00:18Z	Jaroslav Henner	Martin Magr	Red Hat OpenStack	Openstack-packstack	CLOSED	x86_64	Linux	Low	2.0 (Folsom)
916615	2013-02-28T13:35:59Z	2022-07-09T06:03:53Z	Kashyap Chamarthy	Nikola Dipanov	Red Hat OpenStack	Openstack-nova	CLOSED	Unspecified	Linux	Unspecified	2.1
916619	2013-02-28T13:45:04Z	2016-04-26T14:37:24Z	Meital Bourvine	RHOS Maint	Red Hat OpenStack	Openstack-glance	CLOSED	Unspecified	Unspecified	Urgent	2.0 (Folsom)
916622	2013-02-28T14:05:09Z	2016-04-26T17:59:59Z	Derek Higgins	Pete Zaitcev	Red Hat OpenStack	Openstack-swift	CLOSED	Unspecified	Unspecified	Medium	2.1
916626	2013-02-28T14:09:52Z	2016-04-26T21:33:26Z	Derek Higgins	RHOS Maint	Red Hat OpenStack	Openstack-swift	CLOSED	Unspecified	Unspecified	High	2.1
916649	2013-02-28T15:25:39Z	2019-09-09T16:45:51Z	jliberma@redhat.com	Attila Fazekas	Red Hat OpenStack	Openstack-nova	CLOSED	Unspecified	Unspecified	Low	2.1
916671	2013-02-28T16:10:16Z	2016-04-26T16:54:44Z	Attila Fazekas	RHOS Maint	Red Hat OpenStack	Openstack-glance	CLOSED	Unspecified	Unspecified	Urgent	2.0 (Folsom)
917059	2013-03-01T15:17:46Z	2022-07-09T06:05:20Z	Attila Fazekas	Eoghan Glynn	Red Hat OpenStack	Openstack-glance	CLOSED	Unspecified	Unspecified	Unspecified	2.0 (Folsom)

(continued)

8.4 Numerical Examples

Table 8.2 (continued)

Bug ID	Opened	Changed	Reporter	Product	Component	Status	Resolution	Hardware	OS	Severity	Version
917073	2013-03-01T15:38:43Z	2014-01-12T23:55:02Z	Rami Vaknin	Derek Higgins	Red Hat OpenStack	Openstack-packstack	CLOSED	Unspecified	Unspecified	Medium	2.0 (Folsom)
917101	2013-03-01T17:27:49Z	2019-09-09T14:39:19Z	Daniel Berrang	Daniel Berrang	Red Hat OpenStack	Openstack-nova	CLOSED	Unspecified	Unspecified	Medium	2.1
917122	2013-03-01T19:06:04Z	2016-04-26T17:20:14Z	Attila Fazekas	Eric Harney	Red Hat OpenStack	Openstack-cinder	CLOSED	Unspecified	Unspecified	Low	2.0 (Folsom)
917125	2013-03-01T19:17:40Z	2016-04-26T23:04:09Z	Attila Fazekas	Eric Harney	Red Hat OpenStack	Openstack-cinder	CLOSED	Unspecified	Unspecified	Unspecified	3
917208	2013-03-02T03:21:22Z	2022-07-09T06:03:36Z	Adam Young	Adam Young	Red Hat OpenStack	Openstack-keystone	CLOSED	Unspecified	Unspecified	Unspecified	2.0 (Folsom)
917266	2013-03-02T12:50:40Z	2016-04-26T13:50:30Z	Attila Fazekas	Eoghan Glynn	Red Hat OpenStack	Openstack-glance	CLOSED	Unspecified	Unspecified	Low	3
917293	2013-03-02T18:44:38Z	2019-09-09T15:37:00Z	Attila Fazekas	RHOS Maint	Red Hat OpenStack	Openstack-nova	CLOSED	Unspecified	Unspecified	Unspecified	2.1
917370	2013-03-03T16:32:42Z	2019-09-09T14:50:05Z	Omri Hochman	Nikola Dipanov	Red Hat OpenStack	Openstack-nova	CLOSED	x86_64	Linux	Urgent	2.0 (Folsom)
917534	2013-03-04T09:59:20Z	2022-07-09T06:03:53Z	Omri Hochman	Lon Hohberger	Red Hat OpenStack	Openstack-nova	CLOSED	x86_64	Linux	Urgent	2.0 (Folsom)
917657	2013-03-04T13:21:03Z	2016-04-26T13:42:03Z	Jaroslav Henner	Pete Zaitcev	Red Hat OpenStack	Openstack-swift	CLOSED	x86_64	Linux	Medium	2.0 (Folsom)
917979	2013-03-05T08:45:37Z	2019-09-10T14:12:03Z	Meital Bourvine	Adam Young	Red Hat OpenStack	Openstack-keystone	CLOSED	Unspecified	Unspecified	Low	2.0 (Folsom)
918113	2013-03-05T14:05:26Z	2019-09-09T13:27:15Z	Omri Hochman	Brent Eagles	Red Hat OpenStack	Openstack-nova	CLOSED	x86_64	Linux	Medium	2.0 (Folsom)
918115	2013-03-05T14:09:03Z	2014-10-30T22:34:36Z	Ohad Basan	RHOS Maint	Red Hat OpenStack	Openstack-packstack	CLOSED	Unspecified	Unspecified	Unspecified	Unspecified

(continued)

Table 8.2 (continued)

Bug ID	Opened	Changed	Reporter	Product	Component	Status	Resolution	Hardware	OS	Severity	Version
918159	2013-03-05T15:43:07Z	2022-07-09T06:03:36Z	Adam Young	Adam Young	Red Hat OpenStack	Openstack-keystone	CLOSED	Unspecified	Unspecified	Unspecified	2.0 (Folsom)
918185	2013-03-05T16:42:41Z	2016-04-27T04:40:58Z	Stephen Gordon	Martin Magr	Red Hat OpenStack	Openstack-packstack	CLOSED	Unspecified	Unspecified	Low	2.1
918304	2013-03-05T20:57:47Z	2013-04-04T18:00:59Z	Stephen Gordon	Martin Magr	Red Hat OpenStack	Openstack-packstack	CLOSED	Unspecified	Unspecified	Unspecified	2.1
918504	2013-03-06T12:12:21Z	2013-03-06T13:45:02Z	Jaroslav Henner	RHOS Maint	Red Hat OpenStack	Openstack-packstack	CLOSED	x86_64	Linux	High	2.0 (Folsom)
918530	2013-03-06T13:15:31Z	2019-09-09T16:26:32Z	Ofer Blaut	Brent Eagles	Red Hat OpenStack	Openstack-nova	CLOSED	Unspecified	Unspecified	Urgent	2.0 (Folsom)
918721	2013-03-06T18:21:52Z	2014-10-07T00:16:10Z	Lon Hohberger	Lon Hohberger	Red Hat OpenStack	Openstack-selinux	CLOSED	Unspecified	Unspecified	Urgent	2.1
918761	2013-03-06T19:58:22Z	2019-09-09T13:21:53Z	Jon Thomas	David Ripton	Red Hat OpenStack	Openstack-nova	CLOSED	Unspecified	Linux	High	2.0 (Folsom)
919046	2013-03-07T13:39:52Z	2019-09-09T13:45:34Z	Rami Vaknin	Russell Bryant	Red Hat OpenStack	Openstack-nova	CLOSED	Unspecified	Unspecified	Medium	2.0 (Folsom)
919071	2013-03-07T14:32:15Z	2013-04-04T18:01:23Z	Nir Magnezi	Derek Higgins	Red Hat OpenStack	Openstack-packstack	CLOSED	Unspecified	Linux	High	2.1
919074	2013-03-07T14:37:23Z	2019-09-09T13:16:30Z	Dan Prince	Lon Hohberger	Red Hat OpenStack	Openstack-nova	CLOSED	x86_64	Linux	Urgent	2.1
919106	2013-03-07T15:35:54Z	2019-09-10T14:08:29Z	Attila Fazekas	Nikola Dipanov	Red Hat OpenStack	Openstack-nova	CLOSED	Unspecified	Unspecified	Unspecified	2.1

(continued)

8.4 Numerical Examples

Table 8.2 (continued)

BugID	Opened	Changed	Reporter	Product	Component	Status	Resolution	Hardware	OS	Severity	Version
919181	2013-03-07T19:26:24Z	2016-04-26T17:14:44Z	Ofer Blaut	Bob Kukura	Red Hat OpenStack	Openstack-neutron	CLOSED	Unspecified	Unspecified	High	2.0 (Folsom)
919439	2013-03-08T13:18:50Z	2019-09-09T14:40:58Z	Attila Fazekas	Nikola Dipanov	Red Hat OpenStack	Openstack-nova	CLOSED	Unspecified	Unspecified	Unspecified	2.1
919443	2013-03-08T13:38:18Z	2019-09-09T13:51:58Z	Attila Fazekas	RHOS Maint	Red Hat OpenStack	Openstack-nova	CLOSED	Unspecified	Unspecified	Unspecified	2.1
919497	2013-03-08T16:28:18Z	2016-04-27T00:13:28Z	Alan Pevec	John Bresnahan	Red Hat OpenStack	Openstack-glance	CLOSED	Unspecified	Unspecified	Urgent	2.0 (Folsom)
919526	2013-03-08T17:53:09Z	2022-07-09T06:06:23Z	Alan Pevec	Alan Pevec	Red Hat OpenStack	Openstack-keystone	CLOSED	Unspecified	Unspecified	High	2.0 (Folsom)
919607	2013-03-08T21:51:10Z	2013-12-19T23:59:26Z	Perry Myers	Giulio Fidente	Red Hat OpenStack	Openstack-packstack	CLOSED	Unspecified	Unspecified	Unspecified	2.0 (Folsom)
919799	2013-03-10T07:36:26Z	2019-09-09T17:09:51Z	Ofer Blaut	RHOS Maint	Red Hat OpenStack	Openstack-nova	CLOSED	Unspecified	Unspecified	Urgent	2.0 (Folsom)
919868	2013-03-10T15:36:04Z	2019-09-09T16:56:20Z	Rami Vaknin	Russell Bryant	Red Hat OpenStack	Openstack-nova	CLOSED	Unspecified	Unspecified	Medium	2.0 (Folsom)
920024	2013-03-11T07:58:10Z	2019-09-09T16:52:36Z	Ofer Blaut	Brent Eagles	Red Hat OpenStack	Openstack-neutron	CLOSED	Unspecified	Unspecified	Low	2.0 (Folsom)
920213	2013-03-11T15:13:39Z	2013-04-04T18:02:51Z	Jaroslav Henner	Derek Higgins	Red Hat OpenStack	Openstack-packstack	CLOSED	x86_64	Linux	Low	2.0 (Folsom)
920230	2013-03-11T15:41:59Z	2013-05-20T09:32:12Z	Jaroslav Henner	Martin Magr	Red Hat OpenStack	Openstack-packstack	CLOSED	x86_64	Linux	High	2.0 (Folsom)
920282	2013-03-11T17:11:23Z	2019-09-09T15:13:51Z	Omri Hochman	Nikola Dipanov	Red Hat OpenStack	Openstack-nova	CLOSED	x86_64	Linux	High	2.0 (Folsom)
920291	2013-03-11T17:41:09Z	2013-05-29T15:06:27Z	Russell Bryant	RHOS Maint	Red Hat OpenStack	Openstack-packstack	CLOSED	Unspecified	Unspecified	Low	2.0 (Folsom)

(continued)

Table 8.2 (continued)

BugID	Opened	Changed	Reporter	Product	Component	Status	Resolution	Hardware	OS	Severity	Version
920312	2013-03-11T19:09:33Z	2016-04-27T02:37:28Z	Eric Harney	Eric Harney	Red Hat OpenStack	Openstack-cinder	CLOSED	Unspecified	Unspecified	Medium	2.0 (Folsom)
920328	2013-03-11T20:28:29Z	2019-09-09T15:20:04Z	seth vidal	Brent Eagles	Red Hat OpenStack	Openstack-nova	CLOSED	Unspecified	Unspecified	Low	2.0 (Folsom)
920366	2013-03-11T21:53:47Z	2019-09-09T14:09:29Z	Eric Harney	Eric Harney	Red Hat OpenStack	Openstack-cinder	CLOSED	Unspecified	Unspecified	Unspecified	2.1
920516	2013-03-12T09:54:28Z	2013-06-11T18:50:22Z	Nir Magnezi	Martin Magr	Red Hat OpenStack	Openstack-packstack	CLOSED	Unspecified	Linux	Medium	2.1
920638	2013-03-12T13:28:58Z	2016-04-26T18:36:29Z	Bob Kukura	Bob Kukura	Red Hat OpenStack	Openstack-neutron	CLOSED	Unspecified	Unspecified	High	2.0 (Folsom)
920704	2013-03-12T15:13:27Z	2019-09-09T13:32:11Z	Brent Eagles	Martin Magr	Red Hat OpenStack	Openstack-packstack	CLOSED	Unspecified	Unspecified	High	2.0 (Folsom)
920738	2013-03-12T16:02:33Z	2013-08-11T16:42:42Z	Eric Harney	RHOS Maint	Red Hat OpenStack	Openstack-selinux	CLOSED	Unspecified	Unspecified	Unspecified	2.1
920867	2013-03-12T22:22:03Z	2016-04-26T21:06:43Z	Russell Bryant	Derek Higgins	Red Hat OpenStack	Openstack-packstack	CLOSED	Unspecified	Unspecified	Low	2.0 (Folsom)
920982	2013-03-13T08:55:08Z	2016-04-26T22:18:54Z	Flavio Percoco	Flavio Percoco	Red Hat OpenStack	Openstack-glance	CLOSED	Unspecified	Unspecified	Unspecified	4
921517	2013-03-14T11:10:01Z	2019-09-09T13:07:22Z	Jiri Stransky	Nikola Dipanov	Red Hat OpenStack	Openstack-nova	CLOSED	Unspecified	Unspecified	Low	2.1
921692	2013-03-14T17:07:16Z	2013-03-15T12:51:54Z	Martin Pavlsek	RHOS Maint	Red Hat OpenStack	Openstack-packstack	CLOSED	x86_64	Unspecified	Unspecified	2.0 (Folsom)
923086	2013-03-19T07:36:09Z	2019-09-09T14:37:23Z	Ofer Blaut	Nikola Dipanov	Red Hat OpenStack	Openstack-nova	CLOSED	Unspecified	Unspecified	Low	2.1
923115	2013-03-19T09:02:27Z	2019-09-09T14:31:06Z	Gary Kotton	David Ripton	Red Hat OpenStack	Openstack-nova	CLOSED	Unspecified	Unspecified	High	2.1

(continued)

Table 8.2 (continued)

BugID	Opened	Changed	Reporter	Product	Component	Status	Resolution	Hardware	OS	Severity	Version
923226	2013-03-19T13:25:27Z	2016-04-26T14:11:46Z	Eoghan Glynn	Flavio Percoco	Red Hat OpenStack	Openstack-glance	CLOSED	Unspecified	Unspecified	Unspecified	2.1
923395	2013-03-19T18:16:22Z	2022-07-09T06:05:29Z	Attila Fazekas	Eoghan Glynn	Red Hat OpenStack	Openstack-glance	CLOSED	Unspecified	Unspecified	Unspecified	2.0 (Folsom)
923426	2013-03-19T19:28:07Z	2013-04-04T18:04:03Z	Lon Hohberger	Lon Hohberger	Red Hat OpenStack	Openstack-selinux	CLOSED	Unspecified	Unspecified	Medium	2.1
923612	2013-03-20T08:08:01Z	2016-04-27T03:59:34Z	Ofer Blaut	Adam Young	Red Hat OpenStack	Openstack-keystone	CLOSED	Unspecified	Unspecified	Medium	2.1
923621	2013-03-20T08:28:39Z	2019-09-09T14:04:29Z	Meital Bourvine	Nikola Dipanov	Red Hat OpenStack	Openstack-nova	CLOSED	x86_64	Linux	High	2.0 (Folsom)
923650	2013-03-20T09:44:51Z	2019-09-10T14:10:16Z	Omri Hochman	Derek Higgins	Red Hat OpenStack	Openstack-packstack	CLOSED	x86_64	Linux	Medium	2.0 (Folsom)
923666	2013-03-20T09:59:23Z	2016-04-27T01:13:10Z	Flavio Percoco	Flavio Percoco	Red Hat OpenStack	Openstack-glance	CLOSED	Unspecified	Unspecified	Medium	2.0 (Folsom)
923912	2013-03-20T17:22:48Z	2016-04-26T19:30:49Z	Pavel Sedl	Alan Pevec	Red Hat OpenStack	Openstack-keystone	CLOSED	Unspecified	Unspecified	Low	2.1
924000	2013-03-20T22:02:29Z	2019-09-09T15:58:53Z	Alvaro Lopez Ortega	Russell Bryant	Red Hat OpenStack	Openstack-nova	CLOSED	Unspecified	Unspecified	High	2.1
924159	2013-03-21T10:11:26Z	2013-04-04T18:04:15Z	Derek Higgins	Derek Higgins	Red Hat OpenStack	Openstack-packstack	CLOSED	x86_64	Linux	Medium	2.1
924336	2013-03-21T15:03:59Z	2016-04-26T14:43:01Z	Nir Magnezi	Martin Magr	Red Hat OpenStack	Openstack-packstack	CLOSED	Unspecified	Linux	High	2.1
924358	2013-03-21T15:41:19Z	2013-04-04T18:04:19Z	Nir Magnezi	Derek Higgins	Red Hat OpenStack	Openstack-packstack	CLOSED	Unspecified	Linux	Medium	2.1
924463	2013-03-21T20:16:19Z	2016-04-26T18:30:35Z	Pete Zaitcev	Pete Zaitcev	Red Hat OpenStack	Openstack-swift	CLOSED	Unspecified	Unspecified	Medium	2.1

(continued)

Table 8.2 (continued)

BugID	Opened	Changed	Reporter	Product	Component	Status	Resolution	Hardware	OS	Severity	Version
924904	2013-03-22T19:15:37Z	2016-04-26T22:18:09Z	Jon Thomas	RHOS Maint	Red Hat OpenStack	Openstack-glance	CLOSED	Unspecified	Unspecified	High	2.0 (Folsom)
926921	2013-03-24T06:40:10Z	2019-09-09T13:58:52Z	Gary Kotton	RHOS Maint	Red Hat OpenStack	Openstack-keystone	CLOSED	Unspecified	Unspecified	Unspecified	2.1
927167	2013-03-25T09:20:20Z	2019-09-09T17:09:21Z	Kashyap Chamarthy	Nikola Dipanov	Red Hat OpenStack	Openstack-nova	CLOSED	Unspecified	Linux	Medium	2.0 (Folsom)
927349	2013-03-25T17:45:01Z	2019-09-09T15:06:20Z	David Hill	Brent Eagles	Red Hat OpenStack	Openstack-nova	CLOSED	x86_64	Linux	High	Unspecified
927381	2013-03-25T19:38:28Z	2016-04-26T17:22:46Z	Lon Hohberger	Flavio Percoco	Red Hat OpenStack	Openstack-glance	CLOSED	Unspecified	Unspecified	Medium	2.1
927423	2013-03-25T21:13:46Z	2016-04-26T17:44:52Z	Lon Hohberger	Pete Zaitcev	Red Hat OpenStack	Openstack-swift	CLOSED	Unspecified	Unspecified	Medium	2.1
927929	2013-03-26T14:21:05Z	2016-04-26T14:23:08Z	Adam Young	Adam Young	Red Hat OpenStack	Openstack-keystone	CLOSED	Unspecified	Unspecified	High	2.1
928040	2013-03-26T18:45:58Z	2016-04-26T13:38:12Z	Ryan O'Hara	Terry Wilson	Red Hat OpenStack	Openstack-packstack	CLOSED	Unspecified	Unspecified	Unspecified	2.0 (Folsom)
928424	2013-03-27T15:56:00Z	2016-04-27T02:16:11Z	Ryan O'Hara	Terry Wilson	Red Hat OpenStack	Openstack-packstack	CLOSED	Unspecified	Unspecified	Unspecified	2.0 (Folsom)
928491	2013-03-27T18:33:34Z	2013-04-30T12:16:51Z	jliberma@redhat.com	RHOS Maint	Red Hat OpenStack	Openstack-packstack	CLOSED	Unspecified	Unspecified	Unspecified	2.1
928757	2013-03-28T11:48:58Z	2019-09-09T17:17:12Z	Nicolas	Brent Eagles	Red Hat OpenStack	Openstack-nova	CLOSED	Unspecified	Unspecified	Unspecified	2.0 (Folsom)

8.4 Numerical Examples

Table 8.3 A part of the numeric values converted from the raw data

Bug ID	Opened	Changed	Reporter	Assignee	Product	Component	Status	Hardware	OS	Severity	Version	Summary
835466	6.884733796	372.7222801	0.000139127	0.00250429	1	0.165236748	1	0.156147104	0.22144414	0.321569355	0.002133284	86
842136	26.17663194	2605.103403	0.000139127	0.011083801	1	0.165236748	1	0.769327088	0.745443584	0.279923944	0.002133284	78
856263	51.05703704	3587.67287	0.001113018	0.001298521	1	0.165236748	1	0.769327088	0.745443584	0.151138524	0.014422854	78
857256	2.343958333	412.7391204	0.004452071	0.00250429	1	0.165236748	1	0.769327088	0.745443584	0.321569355	0.002133284	45
857290	0.166446759	3585.163252	4.64E-05	0.00250429	1	0.023651625	1	0.156147104	0.22144414	0.321569355	0.014422854	65
861175	13.61262731	1302.458877	0.002457914	0.004869452	1	0.165236748	1	0.073134536	0.22144414	0.279923944	0.002133284	61
861492	1.078310185	1301.380567	0.002457914	0.004869452	1	0.165236748	1	0.156147104	0.22144414	0.151138524	0.002133284	57
861508	0.062627315	1306.001968	0.002457914	0.080044521	1	0.165236748	1	0.769327088	0.745443584	0.279923944	0.002133284	123
861516	0.020706019	278.2066319	0.002457914	0.080044521	1	0.165236748	1	0.769327088	0.745443584	0.151138524	0.002133284	40
861943	2.603657407	1303.42941	0.006863609	0.003199926	1	0.023651625	1	0.769327088	0.22144414	0.11213653	0.002133284	25
862322	1.142314815	2533.933438	0.002457914	0.00380281	1	0.023651625	1	0.769327088	0.745443584	0.151138524	0.002133284	34
865314	8.685636574	2524.376493	0.001530399	0.006585355	1	0.165236748	1	0.769327088	0.745443584	0.279923944	0.066085424	87
865316	0.003449074	2524.203681	0.001530399	0.004869452	1	0.165236748	1	0.769327088	0.745443584	0.279923944	0.002133284	91
865335	0.032974537	2612.062188	0.001530399	0.004266568	1	0.165236748	1	0.769327088	0.745443584	0.11213653	0.034225293	46
865336	0.001388889	1293.632928	0.001530399	0.001159393	1	0.165236748	1	0.769327088	0.745443584	0.321569355	0.002133284	108
865337	0.001736111	2594.151516	0.001530399	0.001298521	1	0.165236748	1	0.769327088	0.745443584	0.151138524	0.0330659	81
865338	0.003634259	1288.822442	0.001530399	0.004869452	1	0.165236748	1	0.769327088	0.745443584	0.279923944	0.066085424	82
865340	0.002141204	265.7305556	0.001530399	0.001298521	1	0.165236748	1	0.769327088	0.745443584	0.279923944	0.002133284	40
865347	0.007789352	3557.869363	0.001530399	0.019570561	1	0.035060057	1	0.769327088	0.745443584	0.321569355	0.002133284	96
865505	0.2553125	1293.111192	0.001530399	0.001298521	1	0.165236748	1	0.769327088	0.745443584	0.279923944	0.002133284	90
865924	1.185555556	117.0103241	0.002457914	0.080044521	1	0.165236748	1	0.769327088	0.745443584	0.151138524	0.002133284	69
866193	1.818703704	262.463206	0.001530399	0.080044521	1	0.165236748	1	0.769327088	0.745443584	0.279923944	0.002133284	100
866199	0.01412037	1285.538819	0.001530399	0.004869452	1	0.165236748	1	0.769327088	0.745443584	0.279923944	0.014422854	50
866451	0.828946759	1289.436505	0.004591198	0.00250429	1	0.023651625	1	0.156147104	0.22144414	0.11213653	0.002133284	50
866681	0.377569444	261.2425694	0.004591198	0.080044521	1	0.023651625	1	0.156147104	0.22144414	0.279923944	0.002133284	102
867029	0.753657407	3552.6139	0.001901405	0.00250429	1	0.023651625	1	0.769327088	0.745443584	0.151138524	0.002133284	50

(continued)

Table 8.3 (continued)

Bug ID	Opened	Changed	Reporter	Assignee	Product	Component	Status	Hardware	OS	Severity	Version	Summary
867035	0.006793981	1283.571852	0.001901405	0.004869452	1	0.165236748	1	0.769327088	0.745443584	0.151138524	0.002133284	50
874316	22.27342593	238.2095023	0.004591198	0.00250429	1	0.165236748	1	0.156147104	0.22144414	0.135231647	0.014422854	98
875938	4.984513889	1256.313912	0.004591198	0.004869452	1	0.165236748	1	0.156147104	0.22144414	0.279923944	0.014422854	41
877343	3.522071759	2488.173171	0.004591198	0.006678106	1	0.010759171	1	0.156147104	0.22144414	0.11213653	0.014422854	58
878229	3.488425926	1254.205451	0.002457914	0.00881139	1	0.02077633	1	0.073134536	0.22144414	0.11213653	0.002133284	40
880980	8.52255787	217.69273151	4.64E-05	0.080044521	1	0.023651625	1	0.769327088	0.745443584	0.151138524	0.014422854	150
880984	0.020520833	217.67221065	0.008672263	0.00250429	1	0.023651625	1	0.769327088	0.22144414	0.279923944	0.014422854	98
881048	0.128587963	61.88740741	0.008672263	0.019570561	1	0.02077633	1	0.769327088	0.22144414	0.321569355	0.014422854	69
881105	0.115868056	217.4277546	0.004591198	0.003199926	1	0.023651625	1	0.156147104	0.22144414	0.11213653	0.014422854	97
881610	0.6490625	2476.246493	0.008672263	0.032834021	1	0.070027362	1	0.769327088	0.22144414	0.279923944	0.014422854	99
881737	0.225972222	1131.043426	0.001113018	0.019570561	1	0.165236748	1	0.769327088	0.745443584	0.151138524	0.002133284	55
881740	0.001539352	1244.009687	0.008672263	0.032834021	1	0.070027362	1	0.769327088	0.22144414	0.321569355	0.014422854	123
881810	0.089201389	3508.594514	0.008672263	0.019570561	1	0.165236748	1	0.769327088	0.22144414	0.321569355	0.014422854	94
882977	3.945231481	3018.192477	0.008672263	0.019570561	1	0.010759171	1	0.769327088	0.745443584	0.151138524	0.014422854	107
883829	1.905972222	3502.734606	0.001901405	0.00250429	1	0.023651625	1	0.769327088	0.745443584	0.151138524	0.014422854	62
883831	0.00181713	6.03E+00	0.001901405	0.00250429	1	0.02077633	1	0.769327088	0.745443584	0.151138524	0.014422854	60
883836	0.001099537	6.033888889	0.001901405	0.00250429	1	0.070027362	1	0.769327088	0.745443584	0.151138524	0.014422854	60
883844	0.002430556	210.6060764	0.001901405	0.00250429	1	0.165236748	1	0.769327088	0.745443584	0.151138524	0.014422854	58
884449	0.890555556	5.141122685	0.008672263	0.00250429	1	0.035060057	1	0.769327088	0.22144414	0.279923944	0.014422854	73
884595	0.059849537	1237.502245	0.001901405	0.000032463	1	0.165236748	1	0.769327088	0.745443584	0.321569355	0.014422854	59
884714	0.191655093	4.889641204	0.008672263	0.00250429	1	0.035060057	1	0.769327088	0.22144414	0.321569355	0.014422854	80
884748	0.028055556	355.9359375	0.008672263	0.001020266	1	0.035060057	1	0.769327088	0.22144414	0.279923944	0.014422854	71
884757	0.00775463	4.853865741	0.008672263	0.00250429	1	0.035060057	1	0.769327088	0.22144414	0.321569355	0.014422854	63
885431	2.812349537	0.02806713	0.008672263	0.00250429	1	0.035060057	1	0.769327088	0.22144414	0.11213653	0.014422854	123

(continued)

8.4 Numerical Examples

Table 8.3 (continued)

Bug ID	Opened	Changed	Reporter	Product	Component	Status	Resolution	Hardware	OS	Severity	Version	Summary
885529	0.544074074	2464.641354	0.001391272	0.00816213	1	0.016602514	1	0.769327088	0.745443584	0.321569355	0.014422854	46
885530	0.01068287	1233.498692	0.001391272	0.00250429	1	0.010759171	1	0.769327088	0.745443584	0.151138524	0.014422854	36
885793	0.619837963	1233.108206	0.008672263	0.00881139	1	0.035060057	1	0.769327088	0.22144414	0.135231647	0.014422854	84
885888	0.246481481	65.83887731	0.000278254	0.00250429	1	0.010759171	1	0.769327088	0.745443584	0.151138524	0.066085424	35
886177	0.7921875	48.74575231	0.001391272	0.019570561	1	0.035060057	1	0.769327088	0.745443584	0.279923944	0.014422854	31
886272	0.236770833	2462.651806	0.001159393	0.011083801	1	0.119000139	1	0.769327088	0.745443584	0.151138524	0.005425961	83
886292	0.041041667	2462.576979	0.001159393	0.004869452	1	0.165236748	1	0.769327088	0.745443584	0.11213653	0.005425961	58
886541	0.595289352	2462.996169	0.001391272	0.019570561	1	0.035060057	1	0.769327088	0.745443584	0.321569355	0.014422854	41
886592	0.081956019	3495.623171	0.008672263	0.019570561	1	0.035060057	1	0.769327088	0.22144414	0.279923944	0.014422854	128
886603	0.012384259	3495.579271	0.008672263	0.019570561	1	0.035060057	1	0.769327088	0.22144414	0.321569355	0.014422854	125
887209	1.747025463	1220.859468	0.007651996	0.080044521	1	0.165236748	1	0.769327088	0.745443584	0.151138524	0.014422854	91
887217	0.015023148	1116.164688	0.001113018	0.00250429	1	0.165236748	1	0.769327088	0.745443584	0.279923944	0.014422854	54
887299	0.206643519	1229.28728	0.001391272	0.019570561	1	0.035060057	1	0.769327088	0.745443584	0.321569355	0.014422854	54
887300	0.002118056	290.9519444	0.001391272	0.080044521	1	0.035060057	1	0.769327088	0.745443584	0.321569355	0.014422854	31
887303	0.00369213	3493.592002	0.004544822	0.00250429	1	0.165236748	1	0.769327088	0.745443584	0.321569355	0.014422854	40
887334	0.060347222	2459.988912	0.002457914	0.004869452	1	0.165236748	1	0.769327088	0.745443584	0.279923944	0.002133284	64
887804	2.772164352	1226.183889	0.001066642	0.032834021	1	0.070027362	1	0.769327088	0.745443584	0.279923944	0.002133284	51
887811	0.005243056	1226.558808	0.001066642	0.000510133	1	0.02077633	1	0.769327088	0.745443584	0.279923944	0.002133284	54
887815	0.002118056	3490.763854	0.001066642	0.00250429	1	0.023651625	1	0.769327088	0.745443584	0.279923944	0.002133284	68
887818	0.002291667	1226.371285	0.001066642	0.006678106	1	0.010759171	1	0.769327088	0.745443584	0.279923944	0.002133284	58
887910	0.143587963	1226.023403	0.004544822	0.080044521	1	0.165236748	1	0.769327088	0.745443584	0.279923944	0.002133284	37
887956	0.062731481	1226.107072	0.001391272	0.00250429	1	0.035060057	1	0.769327088	0.745443584	0.151138524	0.014422854	42
887968	0.019872685	3490.574491	0.004544822	0.001298521	1	0.165236748	1	0.769327088	0.745443584	0.279923944	0.002133284	70
888066	0.239502315	898.9406134	0.004452071	0.003199926	1	0.023651625	1	0.769327088	0.745443584	0.151138524	0.005425961	70
888077	0.022673611	3490.277222	0.004452071	0.000510133	1	0.02077633	1	0.769327088	0.745443584	0.279923944	0.005425961	47
888082	0.01681713	198.1133102	0.004452071	0.080044521	1	0.02077633	1	0.769327088	0.745443584	0.151138524	0.005425961	35
888091	0.015046296	1225.76691	0.000510133	0.000510133	1	0.02077633	1	0.769327088	0.745443584	0.11213653	0.014422854	45

(continued)

Table 8.3 (continued)

Bug ID	Opened	Changed	Reporter	Assignee	Product	Component	Status	Hardware	OS	Severity	Version	Summary
888098	0.011979167	198.0862847	0.004452071	0.080044521	1	0.001947781	1	0.769327088	0.745443584	0.151138524	0.005425961	38
888113	0.062037037	285.9884028	0.000510133	0.000510133	1	0.02077633	1	0.769327088	0.745443584	0.11213653	0.014422854	49
888241	0.37462963	1225.153634	0.003663683	0.00250429	1	0.035060057	1	0.073134536	0.22144414	0.321569355	0.005425961	84
888244	0.018229167	197.6314005	0.002550665	0.003199926	1	0.023651625	1	0.156147104	0.22144414	0.11213653	0.014422854	19
888281	0.045486111	1225.389606	0.001530399	0.00250429	1	0.035060057	1	0.769327088	0.745443584	0.321569355	0.014422854	58
888298	0.016134259	1225.274097	0.001530399	0.080044521	1	0.035060057	1	0.769327088	0.745443584	0.279923944	0.014422854	78
888303	0.003206019	2457.035336	0.001530399	0.00250429	1	0.035060057	1	0.769327088	0.745443584	0.279923944	0.014422854	76
888328	0.025717593	3489.675567	0.002550665	0.00250429	1	0.023651625	1	0.769327088	0.745443584	0.151138524	0.014422854	45
888332	0.008194444	1225.234792	0.001530399	0.00250429	1	0.035060057	1	0.769327088	0.745443584	0.279923944	0.014422854	76
888345	0.017916667	1225.074109	0.001530399	0.00250429	1	0.035060057	1	0.769327088	0.745443584	0.135231647	0.014422854	88
888357	0.009143519	1225.267546	0.000927515	0.019570561	1	0.035060057	1	0.769327088	0.745443584	0.279923944	0.014422854	72
888360	0.003854167	2456.967975	0.000927515	0.00881139	1	0.035060057	1	0.769327088	0.745443584	0.279923944	0.014422854	32
888376	0.011377315	1225.363218	0.000927515	0.019570561	1	0.035060057	1	0.769327088	0.745443584	0.279923944	0.014422854	74
888408	0.055891204	197.4350926	0.002550665	0.080044521	1	0.02077633	1	0.769327088	0.745443584	0.151138524	0.014422854	33
888444	0.060613426	1225.059965	0.000927515	0.019570561	1	0.035060057	1	0.769327088	0.745443584	0.151138524	0.014422854	56
888461	0.017071759	1220.444248	0.004452071	0.004869452	1	0.165236748	1	0.769327088	0.745443584	0.151138524	0.005425961	53
888512	0.060520833	1225.38206	0.001530399	0.00250429	1	0.035060057	1	0.769327088	0.745443584	0.279923944	0.014422854	74
888514	0.001736111	2456.762384	0.001530399	0.00250429	1	0.035060057	1	0.769327088	0.745443584	0.11213653	0.014422854	81
888537	0.035775463	285.2229051	0.000510133	0.080044521	1	0.02077633	1	0.769327088	0.745443584	0.151138524	0.014422854	31
888542	0.007777778	285.2151273	0.000510133	0.030607986	1	0.02077633	1	0.769327088	0.745443584	0.279923944	0.014422854	91
888545	0.007430556	285.2076968	0.000510133	0.030607986	1	0.02077633	1	0.769327088	0.745443584	0.279923944	0.014422854	90
888575	0.048668981	3489.338056	0.004452071	0.003199926	1	0.023651625	1	0.769327088	0.745443584	0.11213653	0.005425961	64
888683	0.4415625	1224.256887	0.008672263	0.080044521	1	0.035060057	1	0.769327088	0.22144414	0.321569355	0.014422854	95

(continued)

8.4 Numerical Examples

Table 8.3 (continued)

Bug ID	Opened	Changed	Reporter	Product	Component	Status	Resolution	Hardware	OS	Severity	Version	Summary
888712	0.047719907	196.7062269	0.003663683	0.080044521	1	0.165236748	1	0.769327088	0.745443584	0.11213653	0.005425961	56
888725	0.033784722	3488.816319	0.001901405	0.019570561	1	0.035060057	1	0.769327088	0.745443584	0.279923944	0.014422854	80
888744	0.040810185	1224.102292	0.00217966	0.00250429	1	0.035060057	1	0.769327088	0.745443584	0.11213653	0.005425961	25
888751	0.01619213	1224.070336	0.00217966	0.019570561	1	0.035060057	1	0.769327088	0.745443584	0.279923944	0.005425961	53
888756	0.0153125	2456.068762	0.001530399	0.00250429	1	0.035060057	1	0.769327088	0.745443584	0.321569355	0.014422854	78
888784	0.040775463	1224.05375	0.004591198	0.080044521	1	0.035060057	1	0.769327088	0.745443584	0.279923944	0.014422854	76
888931	0.263217593	2455.7636	0.001530399	0.016277883	1	0.165236748	1	0.769327088	0.745443584	0.321569355	0.014422854	67
888983	0.150717593	2454.676088	0.001159393	0.080044521	1	0.165236748	1	0.769327088	0.745443584	0.11213653	0.005425961	54
889199	0.595324074	1223.238738	0.008672263	0.019570561	1	0.035060057	1	0.769327088	0.22144414	0.279923944	0.014422854	119
889269	0.121076389	1223.322176	0.008672263	0.00250429	1	0.035060057	1	0.769327088	0.22144414	0.135231647	0.014422854	76
889336	0.172280093	1375.69485	0.000371006	0.000510133	1	0.02077633	1	0.769327088	0.745443584	0.279923944	0.014561981	69
889756	2.572303241	7.108865741	0.008672263	0.080044521	1	0.035060057	1	0.769327088	0.22144414	0.279923944	0.014422854	95
889764	0.060034722	1220.41765	0.000788388	0.080044521	1	0.165236748	1	0.769327088	0.745443584	0.11213653	0.014422854	85
889766	0.009143519	1220.575995	0.000788388	0.030607986	1	0.02077633	1	0.769327088	0.745443584	0.151138524	0.014422854	87
889782	0.087395833	2451.997257	0.000788388	0.00816213	1	0.016602514	1	0.769327088	0.745443584	0.321569355	0.014422854	130
889868	0.141701389	3484.519167	4.64E-05	0.00032463	1	0.165236748	1	0.769327088	0.745443584	0.321569355	0.014422854	54
890058	1.0753125	1218.969583	0.001530399	0.00081139	1	0.035060057	1	0.769327088	0.745443584	0.11213653	0.014422854	95
890168	0.72274213	9.152372685	0.008672263	0.080044521	1	0.070027362	1	0.769327088	0.22144414	0.279923944	0.014422854	58
890175	0.051469907	1218.352292	0.008672263	0.019570561	1	0.035060057	1	0.769327088	0.22144414	0.321569355	0.014422854	76
890178	0.034849537	51.14136574	0.004544822	0.080044521	1	0.070027362	1	0.769327088	0.745443584	0.279923944	0.002133284	38
890180	0.002141204	184.9842477	0.004544822	0.080044521	1	0.02077633	1	0.769327088	0.745443584	0.279923944	0.002133284	38

(continued)

Table 8.3 (continued)

Bug ID	Opened	Changed	Reporter	Product	Component	Status	Resolution	Hardware	OS	Severity	Version	Summary
890181	0.001701389	621.6060532	0.004544822	0.080044521	1	0.023651625	1	0.769327088	0.745443584	0.279923944	0.002133284	40
890182	0.000717593	621.6053356	0.004544822	0.001113018	1	0.165236748	1	0.769327088	0.745443584	0.11213653	0.002133284	36
890183	0.001087963	1209.649954	0.004544822	0.019570561	1	0.035060057	1	0.769327088	0.745443584	0.279923944	0.002133284	40
890184	0.000775463	1218.092928	0.004544822	0.005054955	1	0.119000139	1	0.769327088	0.745443584	0.279923944	0.002133284	38
890185	0.00087963	190.4903588	0.004544822	0.080044521	1	0.010759171	1	0.769327088	0.745443584	0.279923944	0.002133284	29
890196	0.091539352	1217.999965	0.004544822	0.000463757	1	0.034271669	1	0.073134536	0.22144414	0.279923944	0.014422854	82
890197	0.000555556	1210.336007	0.004544822	0.001530399	1	0.034271669	1	0.073134536	0.22144414	0.279923944	0.014422854	110
890198	0.001296296	891.1835764	0.004544822	0.000139127	1	0.018735797	1	0.073134536	0.22144414	0.279923944	0.014422854	59
890200	0.013229167	1217.835347	0.004544822	0.032834021	1	0.070027362	1	0.073134536	0.22144414	0.279923944	0.014422854	51
890202	0.004560185	609.0265741	0.004544822	0.032834021	1	0.070027362	1	0.073134536	0.22144414	0.279923944	0.014422854	38
890206	0.004953704	190.3742245	0.004544822	0.00881139	1	0.02077633	1	0.073134536	0.22144414	0.279923944	0.014422854	58
890207	0.000740741	154.8775463	0.004544822	0.00881139	1	0.02077633	1	0.073134536	0.22144414	0.279923944	0.014422854	57
890208	0.000439815	650.2610185	0.004544822	0.080044521	1	0.02077633	1	0.073134536	0.22144414	0.279923944	0.014422854	41
890209	0.000613426	2448.864259	0.004544822	0.000510133	1	0.02077633	1	0.073134536	0.22144414	0.279923944	0.0330659	36
890210	0.00056713	363.889456	0.004544822	0.080044521	1	0.02077633	1	0.073134536	0.22144414	0.279923944	0.014422854	58
890211	0.001863426	1217.829144	0.004544822	0.00881139	1	0.02077633	1	0.073134536	0.22144414	0.279923944	0.0330659	100
890295	0.616678241	3481.881204	0.001530399	0.019570561	1	0.035060057	1	0.769327088	0.745443584	0.279923944	0.014422854	121
890330	0.161099537	189.5922222	0.004591198	0.080044521	1	0.165236748	1	0.769327088	0.745443584	0.279923944	0.014422854	89
890354	0.134039352	189.4581829	0.004591198	0.019570561	1	0.035060057	1	0.769327088	0.745443584	0.151138524	0.014422854	82
890512	0.862164352	3480.728912	0.004591198	0.011083801	1	0.165236748	1	0.769327088	0.745443584	0.321569355	0.014422854	44
890521	0.036608796	14.32550926	0.008672263	0.080044521	1	0.035060057	1	0.769327088	0.22144414	0.279923944	0.014422854	134
890527	0.049710648	1216.057639	0.004591198	0.080044521	1	0.119000139	1	0.769327088	0.745443584	0.11213653	0.014422854	50
890539	0.047893519	1215.896065	0.001530399	0.030607986	1	0.02077633	1	0.769327088	0.745443584	0.11213653	0.014422854	131

(continued)

8.4 Numerical Examples

Table 8.3 (continued)

Bug ID	Opened	Changed	Reporter	Product	Component	Status	Resolution	Hardware	OS	Severity	Version	Summary
890822	2.901458333	1213.106308	0.008672263	0.019570561	1	0.035060057	1	0.769327088	0.22144414	0.321569355	0.014422854	94
890841	0.094525463	1213.208785	0.008672263	0.00250429	1	0.035060057	1	0.769327088	0.22144414	0.321569355	0.014422854	93
890968	0.936793981	2442.950567	0.004591198	0.000032463	1	0.165236748	1	0.769327088	0.745443584	0.279923944	0.014422854	43
890978	0.023865741	2443.032882	0.004591198	0.000032463	1	0.165236748	1	0.769327088	0.745443584	0.279923944	0.014422854	66
890994	0.02775463	1212.422488	0.004591198	0.004869452	1	0.165236748	1	0.769327088	0.745443584	0.321569355	0.014422854	93
891051	0.876226852	1211.059676	0.008672263	0.004452071	1	0.165236748	1	0.769327088	0.22144414	0.11213653	0.014422854	153

(continued)

Table 8.3 (continued)

Bug ID	Opened	Changed	Reporter	Product	Component	Status	Resolution	Hardware	OS	Severity	Version	Summary
891347	1.177835648	3474.557095	0.004544822	0.003292677	1	0.165236748	1	0.769327088	0.745443584	0.151138524	0.014422854	106
891349	0.001921296	3474.563229	0.000278254	0.004869452	1	0.165236748	1	0.769327088	0.745443584	0.151138524	0.014422854	29
891350	0.000902778	1096.909803	0.004544822	0.003292677	1	0.165236748	1	0.769327088	0.745443584	0.151138524	0.014422854	107
891408	0.166180556	1209.928403	0.004544822	0.004869452	1	0.165236748	1	0.769327088	0.745443584	0.279923944	0.014422854	106
891417	0.030763889	12.05415509	0.004544822	0.00250429	1	0.035060057	1	0.769327088	0.745443584	0.321569355	0.014422854	75
891420	0.001759259	3474.354977	0.004544822	0.003292677	1	0.165236748	1	0.769327088	0.745443584	0.11213653	0.014422854	76
891422	0.008923611	2440.800081	0.004544822	0.004869452	1	0.165236748	1	0.769327088	0.745443584	0.11213653	0.014422854	100
891509	0.452638889	1209.235544	0.006863609	0.032834021	1	0.070027362	1	0.769327088	0.745443584	0.279923944	0.014422854	89
891700	0.391689815	27.08983796	0.000032463	0.000032463	1	0.035060057	1	0.769327088	0.745443584	0.151138524	0.066085424	43
892242	2.431203704	0.033576389	0.004544822	0.080044521	1	0.035060057	1	0.769327088	0.745443584	0.11213653	0.002133284	102
892247	0.035891204	3471.044653	0.004544822	0.019570561	1	0.035060057	1	0.769327088	0.745443584	0.11213653	0.014422854	43
892249	0.011643519	4.64681713	0.004544822	0.080044521	1	0.035060057	1	0.769327088	0.745443584	0.279923944	0.014422854	75
892300	0.306770833	178.5837037	0.001530399	0.019570561	1	0.02077633	1	0.769327088	0.745443584	0.279923944	0.014422854	40
892318	0.161122685	22.76454861	0.004544822	0.003292677	1	0.035060057	1	0.769327088	0.745443584	0.279923944	0.014422854	104
892327	0.031631944	1206.035093	0.004544822	0.003199926	1	0.023651625	1	0.769327088	0.745443584	0.151138524	0.014422854	78
892602	0.74630787	2506.055706	0.000278254	0.001020266	1	0.165236748	1	0.769327088	0.745443584	0.279923944	0.005425961	68
892686	0.157729167	3469.600706	0.004544822	0.032834021	1	0.070027362	1	0.769327088	0.745443584	0.279923944	0.014422854	76
892942	0.776712963	3468.866285	0.003988313	0.019570561	1	0.035060057	1	0.156147104	0.22144414	0.321569355	0.014422854	126
893007	0.103298611	2436.056875	0.003988313	0.019570561	1	0.035060057	1	0.156147104	0.22144414	0.279923944	0.014422854	92
893045	0.050555556	1204.1875	0.001530399	0.00088139	1	0.02077633	1	0.769327088	0.745443584	0.279923944	0.014422854	116
893072	0.011516204	1203.991863	0.001530399	0.00088139	1	0.02077633	1	0.769327088	0.745443584	0.279923944	0.014422854	74
893100	0.043425926	2434.996782	0.008672263	0.00250429	1	0.165236748	1	0.769327088	0.22144414	0.279923944	0.014422854	101
893107	0.008472222	3468.617338	0.003988313	0.019570561	1	0.035060057	1	0.156147104	0.22144414	0.279923944	0.014422854	118
893374	0.746296296	2434.185428	0.001066642	0.004869452	1	0.165236748	1	0.769327088	0.745443584	0.151138524	0.014561981	40
893379	0.008981481	2434.288426	0.001066642	0.004869452	1	0.165236748	1	0.769327088	0.745443584	0.151138524	0.066085424	36

(continued)

8.4 Numerical Examples

Table 8.3 (continued)

Bug ID	Opened	Changed	Reporter	Product	Component	Status	Resolution	Hardware	OS	Severity	Version	Summary
893449	0.061041667	2434.234896	0.002550665	0.004869452	1	0.165236748	1	0.769327088	0.745443584	0.321569355	0.066085424	48
893669	0.220729167	3.61E+01	0.00032463	0.00881139	1	0.035060057	1	0.769327088	0.745443584	0.321569355	0.066085424	49
893934	0.737268519	35.34394676	0.008672263	0.00881139	1	0.035060057	1	0.769327088	0.22144414	0.279923944	0.014422854	80
893938	0.009027778	2433.243067	0.006863609	0.00064926	1	0.165236748	1	0.769327088	0.745443584	0.279923944	0.014422854	27
894460	1.406377315	874.071169	0.004544822	0.008347633	1	0.035060057	1	0.769327088	0.745443584	0.11213653	0.014422854	75
894712	1.665752315	25.49131944	0.003988313	0.080044521	1	0.035060057	1	0.156147104	0.22144414	0.135231647	0.014422854	115
894715	0.010324074	2430.143218	0.002550665	0.080044521	1	0.165236748	1	0.769327088	0.745443584	0.135231647	0.014422854	58
894733	0.116689815	3463.635313	0.001530399	0.019570561	1	0.035060057	1	0.769327088	0.745443584	0.279923944	0.014422854	123
894789	0.17869213	2341.518137	0.004544822	0.080044521	1	0.119000139	1	0.073134536	0.22144414	0.321569355	0.014422854	71
894794	0.002349537	1199.327338	0.004544822	0.080044521	1	0.023651625	1	0.073134536	0.22144414	0.279923944	0.014422854	79
894795	0.000740741	1198.900428	0.004544822	0.080044521	1	0.023651625	1	0.073134536	0.22144414	0.279923944	0.014422854	96
894796	0.00037037	1198.780984	0.004544822	0.080044521	1	0.023651625	1	0.073134536	0.22144414	0.279923944	0.014422854	56
894812	0.008888889	2429.889155	0.004544822	0.001113018	1	0.165236748	1	0.073134536	0.22144414	0.279923944	0.014422854	59
894813	0.001319444	2429.782986	0.004544822	0.001159393	1	0.165236748	1	0.073134536	0.22144414	0.11213653	0.014422854	91
894814	0.000798611	2429.819572	0.004544822	0.080044521	1	0.165236748	1	0.073134536	0.22144414	0.279923944	0.014422854	60
894815	0.000787037	1198.922199	0.004544822	0.001530399	1	0.119000139	1	0.073134536	0.22144414	0.279923944	0.014422854	44
894820	0.002766204	2429.857627	0.004544822	0.004869452	1	0.165236748	1	0.073134536	0.22144414	0.279923944	0.014422854	78
894824	0.001516204	2429.754178	0.004544822	4.64E-05	1	0.165236748	1	0.073134536	0.22144414	0.279923944	0.014561981	49
894826	0.000856481	16.8424537	0.004544822	0.080044521	1	0.035060057	1	0.073134536	0.22144414	0.279923944	0.014422854	42
894827	0.000798611	101.8799421	0.004544822	0.080044521	1	0.035060057	1	0.073134536	0.22144414	0.279923944	0.014422854	61
894828	0.001261574	148.9521991	0.004544822	0.005054955	1	0.035060057	1	0.073134536	0.22144414	0.279923944	0.014422854	28
894841	0.010555556	1199.024653	0.004544822	0.009414274	1	0.119000139	1	0.073134536	0.22144414	0.279923944	0.014422854	38
894843	0.000706019	1198.736586	0.004544822	0.080044521	1	0.119000139	1	0.073134536	0.22144414	0.279923944	0.014422854	52
894845	0.0015625	2429.77441	0.004544822	0.00032463	1	0.165236748	1	0.073134536	0.22144414	0.279923944	0.014422854	90
894860	0.005601852	2429.725729	0.004544822	0.001159393	1	0.165236748	1	0.073134536	0.22144414	0.11213653	0.014422854	70

(continued)

Table 8.3 (continued)

Bug ID	Opened	Changed	Reporter	Product	Component	Status	Resolution	Hardware	OS	Severity	Version	Summary
894861	0.000405093	2429.824421	0.004544822	0.080044521	1	0.165236748	1	0.073134536	0.22144414	0.279923944	0.014422854	62
894862	0.000462963	872.058125	0.004544822	0.00816213	1	0.016602514	1	0.073134536	0.22144414	0.279923944	0.014422854	77
894892	0.10875	2429.637708	0.004544822	0.001298521	1	0.165236748	1	0.073134536	0.22144414	0.279923944	0.014422854	90
894917	0.109733796	2429.579734	0.004544822	0.001113018	1	0.165236748	1	0.073134536	0.22144414	0.279923944	0.014422854	44
894924	0.041863426	1199.114329	0.004544822	0.000556509	1	0.119000139	1	0.073134536	0.22144414	0.279923944	0.014422854	38
894925	0.001365741	1198.632859	0.004544822	0.003199926	1	0.023651625	1	0.073134536	0.22144414	0.279923944	0.014422854	27
895042	0.351226852	1198.592361	0.003988313	0.009182396	1	0.035060057	1	0.156147104	0.22144414	0.11213653	0.014422854	134
895116	0.195983796	363.3356134	0.002550665	0.00250429	1	0.035060057	1	0.769327088	0.745443584	0.135231647	0.014422854	67
895145	0.059652778	14.74568287	0.007327366	0.019570561	1	0.035060057	1	0.769327088	0.745443584	0.151138524	0.005425961	59
895251	0.219791667	1197.677894	0.007327366	0.019570561	1	0.035060057	1	0.769327088	0.745443584	0.151138524	0.005425961	53
895588	0.707453704	2428.028009	0.007327366	0.004452071	1	0.165236748	1	0.769327088	0.745443584	0.151138524	0.014561981	48
895804	0.471990741	1196.485683	4.64E-05	0.019570561	1	0.035060057	1	0.769327088	0.745443584	0.151138524	0.066085424	74
896085	0.510925926	2427.017037	0.002550665	0.001113018	1	0.165236748	1	0.769327088	0.745443584	0.11213653	0.014422854	74
896107	0.044259259	12.79133102	0.008672263	0.019570561	1	0.035060057	1	0.769327088	0.22144414	0.135231647	0.014422854	56
896115	0.013136574	2426.997153	0.000510133	0.000510133	1	0.165236748	1	0.769327088	0.745443584	0.151138524	0.066085424	48
896156	0.085486111	1.27E+01	0.00032463	0.019570561	1	0.035060057	1	0.769327088	0.745443584	0.151138524	0.066085424	47
896181	0.042106481	1.27E+01	0.007327366	0.019570561	1	0.035060057	1	0.769327088	0.745443584	0.151138524	0.005425961	64
896188	0.017766204	12.63292824	0.004591198	0.019570561	1	0.035060057	1	0.769327088	0.745443584	0.279923944	0.014422854	62
896236	0.071863426	103.6078819	0.007327366	0.080044521	1	0.165236748	1	0.769327088	0.745443584	0.151138524	0.005425961	102
896489	0.609340278	2426.11456	0.002550665	0.080044521	1	0.035060057	1	0.769327088	0.745443584	0.11213653	0.014422854	86
896494	0.013055556	11.93873843	0.008672263	0.019570561	1	0.035060057	1	0.769327088	0.22144414	0.321569355	0.014422854	92
896586	0.103240741	2425.948356	0.002550665	0.004869452	1	0.165236748	1	0.769327088	0.745443584	0.279923944	0.014422854	71
896618	0.033055556	3459.602697	0.007327366	0.019570561	1	0.035060057	1	0.769327088	0.745443584	0.321569355	0.014422854	56
896619	0.005081019	2425.922465	0.007373742	0.004869452	1	0.165236748	1	0.769327088	0.745443584	0.151138524	0.005425961	41
896642	0.022141204	945.3323264	0.008672263	0.019570561	1	0.035060057	1	0.769327088	0.22144414	0.135231647	0.014422854	95

(continued)

8.4 Numerical Examples

Table 8.3 (continued)

Bug ID	Opened	Changed	Reporter	Product	Component	Status	Resolution	Hardware	OS	Severity	Version	Summary
896706	0.135277778	2425.829398	0.004544822	0.00816213	1	0.016602514	1	0.156147104	0.22144414	0.135231647	0.014422854	95
901345	0.278090278	11.3619213	0.0001391272	0.019570561	1	0.035060057	1	0.769327088	0.745443584	0.279923944	0.014422854	51
901955	2.205601852	1192.744132	0.001530399	0.080044521	1	0.023651625	1	0.769327088	0.745443584	0.279923944	0.014422854	28
901968	0.099513889	2423.263438	0.002550665	0.011083801	1	0.165236748	1	0.769327088	0.745443584	0.321569355	0.014422854	66
902020	0.157615741	2424.024167	0.004591198	0.00881139	1	0.035060057	1	0.769327088	0.745443584	0.135231647	0.014422854	67
902293	0.863125	2422.125428	0.00097389	0.030607986	1	0.165236748	1	0.769327088	0.22144414	0.151138524	0.005425961	43
902370	0.163854167	119.7691088	0.008672263	0.080044521	1	0.035060057	1	0.769327088	0.745443584	0.321569355	0.005425961	100
902409	0.0607117593	3455.601782	0.002550665	0.011083801	1	0.165236748	1	0.769327088	0.745443584	0.279923944	0.014422854	50
902485	0.136076389	23.97633102	0.007373742	0.019570561	1	0.035060057	1	0.769327088	0.745443584	0.321569355	0.005425961	51
902546	0.137476852	1190.986782	0.007327366	0.0000139127	1	0.035060057	1	0.769327088	0.745443584	0.11213653	0.014561981	38
902703	0.490752315	23.34813657	0.002550665	0.019570561	1	0.035060057	1	0.769327088	0.745443584	0.11213653	0.014422854	27
903095	0.888078704	110.1352083	0.004591198	0.080044521	1	0.035060057	1	0.769327088	0.745443584	0.11213653	0.014422854	67
903124	0.0834375	2420.18787	0.004591198	0.004869452	1	0.165236748	1	0.769327088	0.745443584	0.279923944	0.014422854	48
903153	0.045069444	22.33158565	0.008672263	0.00881139	1	0.035060057	1	0.769327088	0.22144414	0.279923944	0.005425961	96
903187	0.088125	3453.729132	0.001484024	0.019570561	1	0.035060057	1	0.769327088	0.745443584	0.11213653	0.066085424	59
903191	0.006365741	22.23712963	0.001484024	0.00881139	1	0.035060057	1	0.769327088	0.745443584	0.279923944	0.066085424	61
903502	0.816979167	3452.919201	0.000278254	0.019570561	1	0.035060057	1	0.769327088	0.745443584	0.321569355	0.005425961	32
903504	0.003923611	2419.216863	0.004591198	0.0001113018	1	0.165236748	1	0.769327088	0.745443584	0.11213653	0.014422854	68
903545	0.044108796	3452.910648	0.001901405	0.019570561	1	0.035060057	1	0.769327088	0.745443584	0.321569355	0.014422854	63
903547	0.002118056	2419.262743	0.004591198	0.0001113018	1	0.165236748	1	0.769327088	0.745443584	0.11213653	0.014422854	77
903568	0.027696759	2419.136389	0.004591198	0.080044521	1	0.165236748	1	0.769327088	0.745443584	0.279923944	0.014422854	75
903671	0.190324074	2419.085417	0.002550665	0.011083801	1	0.165236748	1	0.769327088	0.745443584	0.279923944	0.014422854	87
903696	0.032337963	2418.911528	0.002550665	0.080044521	1	0.165236748	1	0.769327088	0.745443584	0.279923944	0.014422854	81
903705	0.011886574	2418.923495	0.002550665	0.080044521	1	0.165236748	1	0.769327088	0.745443584	0.279923944	0.014422854	66
903719	0.021296296	1188.24125	0.007373742	0.00250429	1	0.035060057	1	0.769327088	0.745443584	0.151138524	0.005425961	41

(continued)

Table 8.3 (continued)

Bug ID	Opened	Changed	Reporter	Product	Component	Status	Resolution	Hardware	OS	Severity	Version	Summary
903813	0.211828704	3452.3739	0.007327366	0.019570561	1	0.035060057	1	0.769327088	0.745443584	0.151138524	0.005425961	42
904274	1.050636574	1186.669491	0.000278254	0.000414274	1	0.119000139	1	0.769327088	0.745443584	0.112213653	0.014422854	78
904669	1.05099537	3450.257222	0.004544822	0.019570561	1	0.035060057	1	0.769327088	0.745443584	0.321569355	0.014422854	97
904671	0.021238426	18.75185185	0.004544822	0.019570561	1	0.035060057	1	0.769327088	0.745443584	0.135231647	0.014422854	92
904712	0.35380787	32.09201389	0.004591198	0.080044521	1	0.035060057	1	0.769327088	0.745443584	0.112213653	0.014422854	87
904713	0.00537037	2416.180278	0.004591198	0.080044521	1	0.165236748	1	0.769327088	0.745443584	0.151138524	0.014422854	87
904730	0.073506944	2416.155359	0.004591198	0.080044521	1	0.165236748	1	0.769327088	0.745443584	0.321569355	0.014422854	62
904751	0.034270833	2416.106574	0.004591198	0.011083801	1	0.165236748	1	0.769327088	0.745443584	0.321569355	0.014422854	32
904800	0.169039352	1185.09706	0.001530399	0.080044521	1	0.070027362	1	0.769327088	0.745443584	0.279923944	0.005425961	62
904920	0.67587963	2415.245671	0.001530399	0.004452071	1	0.165236748	1	0.769327088	0.745443584	0.112213653	0.005425961	70
905081	0.2779166667	3448.700833	0.002133284	0.019570561	1	0.035060057	1	0.769327088	0.745443584	0.151138524	0.014422854	80
905113	0.052233796	3448.598981	0.004452071	0.004452071	1	0.165236748	1	0.769327088	0.745443584	0.151138524	0.005425961	33
905114	0.0021875	2414.989688	0.004452071	0.032834021	1	0.070027362	1	0.769327088	0.745443584	0.279923944	0.005425961	35
905368	0.734398148	3447.87316	0.001391272	0.019570561	1	0.035060057	1	0.769327088	0.745443584	0.321569355	0.014422854	36
905437	0.142569444	2414.066331	0.004591198	0.001113018	1	0.165236748	1	0.769327088	0.745443584	0.112213653	0.014422854	74
905516	0.084085648	3447.677199	0.003988313	0.019570561	1	0.035060057	1	0.156147104	0.22144414	0.321569355	0.014422854	97
905737	0.538125	3447.092813	0.004544822	0.019570561	1	0.035060057	1	0.769327088	0.745443584	0.279923944	0.014422854	84
905842	0.255590278	3446.852292	0.002550665	0.019570561	1	0.035060057	1	0.769327088	0.745443584	0.321569355	0.014422854	73
905931	0.167592593	1182.105417	0.004591198	0.003199926	1	0.023651625	1	0.769327088	0.745443584	0.112213653	0.014422854	118
906006	0.073703704	3446.642465	0.007327366	0.019570561	1	0.035060057	1	0.769327088	0.745443584	0.112213653	0.005425961	76
906038	0.040347222	932.3271528	0.008672263	0.00881139	1	0.035060057	1	0.769327088	0.22144414	0.279923944	0.005425961	133
906298	0.806666667	2413.088854	0.004591198	0.003292677	1	0.001947781	1	0.156147104	0.22144414	0.112213653	0.014422854	59
906410	0.148217593	3445.602384	0.001484024	0.019570561	1	0.035060057	1	0.769327088	0.745443584	0.151138524	0.066085424	46
906471	0.060115741	130.9331597	0.001391272	0.00250429	1	0.035060057	1	0.769327088	0.745443584	0.321569355	0.014422854	54
906783	0.854282407	3444.691296	0.000788388	0.080044521	1	0.165236748	1	0.769327088	0.745443584	0.279923944	0.014422854	69
907084	1.800335648	1178.46728	0.004591198	0.003199926	1	0.023651625	1	0.769327088	0.745443584	0.321569355	0.014422854	113
907088	0.035925926	2409.281343	0.002550665	0.080044521	1	0.165236748	1	0.769327088	0.745443584	0.321569355	0.014422854	61

(continued)

8.4 Numerical Examples

Table 8.3 (continued)

Bug ID	Opened	Changed	Reporter	Product	Component	Status	Resolution	Hardware	OS	Severity	Version	Summary
907178	0.255231481	3442.600197	0.002550665	0.011083801	1	0.165236748	1	0.769327088	0.745443584	0.11213653	0.014422854	60
907259	0.368275463	238.0608102	0.001159393	0.000510133	1	0.02077633	1	0.769327088	0.745443584	0.151138524	0.005425961	69
907367	0.328599537	1177.808148	0.000783388	0.080044521	1	0.035060057	1	0.769327088	0.745443584	0.151138524	0.014422854	48
907529	0.317951389	318.3263657	0.001530399	0.013402588	1	0.035060057	1	0.769327088	0.745443584	0.279923944	0.005425961	135
907624	0.191400463	3441.389375	0.007327366	0.019570561	1	0.035060057	1	0.769327088	0.745443584	0.11213653	0.005425961	46
907737	0.438877315	3440.949977	0.000788388	0.011083801	1	0.035060057	1	0.769327088	0.745443584	0.11213653	0.014422854	33
908039	0.424143519	2406.941412	0.001530399	0.003199926	1	0.165236748	1	0.769327088	0.745443584	0.279923944	0.005425961	75
908355	0.85087963	1175.142743	0.002133284	0.001113018	1	0.023651625	1	0.769327088	0.745443584	0.321569355	0.014422854	51
908370	0.029131944	2406.047083	0.004591198	0.00032463	1	0.165236748	1	0.769327088	0.745443584	0.11213653	0.014422854	68
908373	0.006157407	3439.644688	0.004591198	0.080044521	1	0.165236748	1	0.769327088	0.745443584	0.279923944	0.014422854	113
908379	0.008252315	1175.257894	0.004591198	0.011083801	1	0.035060057	1	0.769327088	0.745443584	0.11213653	0.014422854	66
908514	0.291770833	2405.738438	0.000556509	0.019570561	1	0.165236748	1	0.769327088	0.745443584	0.279923944	0.014422854	72
908695	0.542430556	3438.851644	0.003988313	0.019570561	1	0.035060057	1	0.156147104	0.22144414	0.321569355	0.014422854	102
908771	0.113391204	3438.698889	0.000788388	0.019570561	1	0.035060057	1	0.769327088	0.745443584	0.11213653	0.014422854	53
908837	0.111006944	3438.588113	0.004591198	0.019570561	1	0.035060057	1	0.156147104	0.22144414	0.279923944	0.014422854	92
908838	0.001666667	3438.586655	0.004591198	0.019570561	1	0.035060057	1	0.156147104	0.22144414	0.279923944	0.014422854	102
908846	0.012164352	3438.574387	0.004591198	0.019570561	1	0.035060057	1	0.156147104	0.22144414	0.279923944	0.014422854	53
908892	0.109166667	1174.078495	0.002457914	0.080044521	1	0.035060057	1	0.769327088	0.22144414	0.11213653	0.014422854	75
908900	0.037152778	3438.428403	0.004591198	0.019570561	1	0.035060057	1	0.156147104	0.22144414	0.321569355	0.014422854	58
908995	0.2228125	3438.18706	0.0001901405	0.00250429	1	0.023651625	1	0.769327088	0.745443584	0.151138524	0.014422854	62
909137	0.426064815	2404.062523	0.001113018	0.003292677	1	0.165236748	1	0.769327088	0.745443584	0.279923944	0.014422854	58
909286	0.123310185	2403.972025	0.001113018	0.003292677	1	0.165236748	1	0.769327088	0.745443584	0.321569355	0.005425961	72
909418	0.1046875	1059.893762	0.001113018	0.032834021	1	0.070027362	1	0.769327088	0.745443584	0.321569355	0.005425961	78

(continued)

Table 8.3 (continued)

Bug ID	Opened	Changed	Reporter	Product	Component	Status	Resolution	Hardware	OS	Severity	Version	Summary
909424	0.023738426	1172.840046	0.002457914	0.00250429	1	0.035060057	1	0.769327088	0.22144414	0.11213653	0.014422854	44
909434	0.030914352	1173.331493	0.000510133	0.000510133	1	0.02077633	1	0.769327088	0.745443584	0.279923944	0.014422854	60
909438	0.033101852	17.72259954	0.004591198	0.00250429	1	0.035060057	1	0.156147104	0.22144414	0.321569355	0.014422854	77
909441	0.005810185	6.071539352	0.000602885	0.019570561	1	0.035060057	1	0.769327088	0.745443584	0.279923944	0.014422854	32
909460	0.041354167	1172.905278	0.007373742	0.032834021	1	0.070027362	1	0.769327088	0.745443584	0.151138524	0.005425961	60
909505	0.16875	1172.914815	0.001391272	0.019570561	1	0.035060057	1	0.769327088	0.745443584	0.279923944	0.014422854	43
909653	1.302037037	2402.269352	0.000788388	0.004452071	1	0.165236748	1	0.769327088	0.745443584	0.321569355	0.014422854	38
909809	0.978032407	2401.411933	0.004591198	0.011083801	1	0.165236748	1	0.769327088	0.745443584	0.279923944	0.014422854	65
909832	0.088912037	2401.321806	0.000788388	0.011083801	1	0.165236748	1	0.769327088	0.745443584	0.151138524	0.014422854	52
910089	0.39599537	3434.51684	0.004591198	0.019570561	1	0.035060057	1	0.156147104	0.22144414	0.321569355	0.014422854	55
910210	0.278043981	3434.199329	0.001391272	0.019570561	1	0.035060057	1	0.769327088	0.745443584	0.321569355	0.014422854	59
910211	0.003194444	3434.182211	0.001391272	0.019570561	1	0.035060057	1	0.769327088	0.745443584	0.321569355	0.014422854	45
910556	0.793506944	50.8847338	0.004591198	0.00250429	1	0.035060057	1	0.156147104	0.22144414	0.321569355	0.014422854	46
910619	0.19125	2399.620949	0.001391272	0.080044521	1	0.165236748	1	0.769327088	0.745443584	0.321569355	0.014422854	61
910727	0.435150463	3432.813032	0.004591198	0.004452071	1	0.165236748	1	0.156147104	0.22144414	0.321569355	0.014422854	104
910747	0.066018519	2398.988229	0.003988313	0.00250429	1	0.165236748	1	0.156147104	0.22144414	0.279923944	0.014422854	86
910806	0.106053241	2398.967176	0.011222928	0.00816213	1	0.165236748	1	0.156147104	0.22144414	0.279923944	0.014422854	85
910818	0.013043981	3432.575833	0.001391272	0.019570561	1	0.035060057	1	0.769327088	0.745443584	0.151138524	0.014422854	42
911005	0.662407407	2398.288414	0.004591198	0.001113018	1	0.165236748	1	0.769327088	0.745443584	0.135231647	0.014422854	73
911103	0.135023148	3431.78353	0.001113018	0.003292677	1	0.165236748	1	0.769327088	0.745443584	0.151138524	0.014422854	78
911149	0.098078704	2398.105602	0.004591198	0.00032463	1	0.165236748	1	0.769327088	0.745443584	0.321569355	0.014422854	82
911190	0.041134259	2397.977465	0.003988313	0.004869452	1	0.165236748	1	0.156147104	0.22144414	0.279923944	0.014422854	75
911225	0.036053241	2398.070162	0.003988313	0.004452071	1	0.165236748	1	0.156147104	0.22144414	0.321569355	0.014422854	63
911568	0.832928241	3659.478021	0.001484024	0.00881139	1	0.02077633	1	0.769327088	0.745443584	0.279923944	0.066085424	69
911626	0.097986111	3430.727222	0.004591198	0.019570561	1	0.035060057	1	0.156147104	0.22144414	0.11213653	0.014422854	29

(continued)

8.4 Numerical Examples

Table 8.3 (continued)

Bug ID	Opened	Changed	Reporter	Product	Component	Status	Resolution	Hardware	OS	Severity	Version	Summary
911653	0.029328704	3430.643461	0.004591198	0.019570561	1	0.035060057	1	0.156147104	0.221444414	0.135231647	0.014422854	34
912006	1.718263889	3428.940127	0.008672263	0.019570561	1	0.035060057	1	0.769327088	0.221444414	0.321569355	0.005425961	72
912078	0.370706019	2394.939421	0.001113018	0.003292677	1	0.165236748	1	0.769327088	0.745443584	0.279923944	0.014422854	44
912143	0.171956019	1164.369664	9.28E-05	0.030607986	1	0.02077633	1	0.769327088	0.745443584	0.151138524	0.014422854	68
912284	0.539791667	3427.855498	0.000788388	0.011083801	1	0.165236748	1	0.769327088	0.745443584	0.321569355	0.014422854	84
912316	0.087118056	2394.178738	0.003988313	0.011083801	1	0.165236748	1	0.156147104	0.221444414	0.321569355	0.014422854	94
912363	0.09943287	2394.03316	0.003988313	0.080044521	1	0.165236748	1	0.156147104	0.221444414	0.279923944	0.014422854	80
912384	0.0315625	3427.629063	0.007373742	0.032834021	1	0.165236748	1	0.769327088	0.745443584	0.151138524	0.005425961	41
912509	0.236261574	1162.862407	0.004452071	0.032834021	1	0.070027362	1	0.769327088	0.745443584	0.279923944	0.005425961	39
912682	0.646041667	1049.103102	4.64E-05	0.003292677	1	0.001947781	1	0.156147104	0.221444414	0.321569355	0.014422854	28
912702	0.039490741	3426.717836	0.004544822	0.019570561	1	0.035060057	1	0.769327088	0.745443584	0.321569355	0.014422854	80
912744	0.077775463	2392.944618	0.004591198	0.001298521	1	0.165236748	1	0.156147104	0.221444414	0.112135365	0.014422854	67
912745	0.003738426	3.43E-03	0.007327366	0.019570561	1	0.035060057	1	0.769327088	0.745443584	0.279923944	0.005425961	45
912768	0.027974537	3426.608634	0.001391272	0.019570561	1	0.035060057	1	0.769327088	0.745443584	0.279923944	0.005425961	46
913195	1.00849537	1160.924005	0.004591198	0.006678106	1	0.010759171	1	0.156147104	0.221444414	0.279923944	0.014422854	39
913197	0.003032407	2391.919063	0.011222928	0.00816213	1	0.016602514	1	0.156147104	0.221444414	0.135231647	0.005425961	60
913293	0.218078704	13.6812037	0.004591198	0.080044521	1	0.035060057	1	0.156147104	0.221444414	0.279923944	0.014422854	73
913513	0.611770833	1047.114722	0.006863609	0.003292677	1	0.001947781	1	0.769327088	0.745443584	0.112135365	0.014422854	66
913561	0.064618056	2391.145058	0.006863609	0.011083801	1	0.165236748	1	0.769327088	0.745443584	0.112135365	0.014422854	73
913613	0.095335648	3424.595903	0.001113018	0.003292677	1	0.165236748	1	0.769327088	0.745443584	0.279923944	0.014422854	69
913616	0.002696759	2390.942292	0.001113018	0.001159393	1	0.165236748	1	0.769327088	0.745443584	0.279923944	0.014561981	58
913619	0.001828704	3680.080729	0.001113018	0.039790382	1	0.165236748	1	0.156147104	0.221444414	0.112135365	0.014561981	49
914648	0.835277778	1159.199606	0.004591198	0.019570561	1	0.035060057	1	0.156147104	0.221444414	0.279923944	0.014422854	48
914736	0.186759259	11.89438657	0.000371006	0.080044521	1	0.035060057	1	0.769327088	0.745443584	0.321569355	0.005425961	37
914759	0.026701389	3423.543507	0.004452071	0.011083801	1	0.165236748	1	0.769327088	0.745443584	0.321569355	0.005425961	78

(continued)

Table 8.3 (continued)

Bug ID	Opened	Changed	Reporter	Product	Component	Status	Resolution	Hardware	OS	Severity	Version	Summary
915274	2.75443287	3420.797025	0.00282892	0.004869452	1	0.165236748	1	0.769327088	0.745443584	0.151138524	0.005425961	124
915365	0.173460648	297.3564352	0.004544822	0.019570561	1	0.035060057	1	0.769327088	0.745443584	0.279923944	0.014422854	71
915382	0.021099537	3420.605	0.007327366	0.019570561	1	0.035060057	1	0.769327088	0.745443584	0.11213653	0.005425961	69
915383	0.001689815	38.08515046	0.007327366	0.019570561	1	0.035060057	1	0.769327088	0.745443584	0.11213653	0.005425961	74
915445	0.114363426	3420.474606	0.001484024	0.00881139	1	0.02077633	1	0.769327088	0.745443584	0.151138524	0.066085424	36
915881	0.939537037	1154.969988	0.004452071	0.019570561	1	0.035060057	1	0.769327088	0.745443584	0.151138524	0.014422854	59
915906	0.055787037	1154.915567	0.004591198	0.00816213	1	0.016602514	1	0.769327088	0.745443584	0.321569355	0.014422854	39
915968	0.120648148	215.1909259	0.000510133	0.080044521	1	0.02077633	1	0.769327088	0.745443584	0.151138524	0.014422854	66
916174	0.65375	3418.703623	0.001066642	0.011083801	1	0.165236748	1	0.769327088	0.745443584	0.151138524	0.005425961	68
916176	0.003368056	3418.700255	0.001113018	0.001298521	1	0.165236748	1	0.769327088	0.745443584	0.321569355	0.005425961	68
916241	0.107511574	3418.597211	0.000371006	0.004869452	1	0.165236748	1	0.769327088	0.745443584	0.135231647	0.005425961	32
916284	0.0921875	1154.067917	0.007651996	0.032834021	1	0.070027362	1	0.769327088	0.745443584	0.321569355	0.005425961	69
916301	0.047696759	1153.888831	0.000881139	0.013402588	1	0.035060057	1	0.769327088	0.745443584	0.279923944	0.005425961	68
916307	0.016365741	0.517662037	0.000602885	0.080044521	1	0.035060057	1	0.769327088	0.745443584	0.279923944	0.014422854	46
916329	0.038125	1153.716111	0.007327366	0.000463757	1	0.001947781	1	0.769327088	0.745443584	0.11213653	0.005425961	44
916558	0.573217593	1040.18897	0.006863609	0.003292677	1	0.001947781	1	0.769327088	0.745443584	0.11213653	0.014422854	85
916567	0.023136574	1040.16537	0.006863609	0.003292677	1	0.001947781	1	0.769327088	0.745443584	0.279923944	0.014422854	68
916605	0.10056713	35.19900463	0.004591198	0.019570561	1	0.035060057	1	0.156147104	0.22144414	0.11213653	0.014422854	87
916615	0.015451389	3417.686042	0.00282892	0.004869452	1	0.165236748	1	0.769327088	0.22144414	0.151138524	0.005425961	66
916619	0.00630787	1153.036343	0.000185503	0.080044521	1	0.02077633	1	0.769327088	0.745443584	0.135231647	0.014422854	25
916622	0.013946759	1153.163079	0.001391272	0.006678106	1	0.010759171	1	0.769327088	0.745443584	0.279923944	0.005425961	74
916626	0.003275463	1153.308032	0.001391272	0.080044521	1	0.010759171	1	0.769327088	0.745443584	0.321569355	0.005425961	64
916649	0.052627315	2384.055694	0.001576775	0.080044521	1	0.165236748	1	0.769327088	0.745443584	0.11213653	0.005425961	64
916671	0.030983796	1153.03088	0.006863609	0.080044521	1	0.02077633	1	0.769327088	0.745443584	0.135231647	0.014422854	32
917059	0.963541667	3416.616366	0.006863609	0.030607986	1	0.02077633	1	0.769327088	0.745443584	0.151138524	0.014422854	79

(continued)

8.4 Numerical Examples

Table 8.3 (continued)

Bug ID	Opened	Changed	Reporter	Product	Component	Status	Resolution	Hardware	OS	Severity	Version	Summary
917073	0.014548611	317.3446644	0.002550665	0.00250429	1	0.035060057	1	0.769327088	0.745443584	0.279923944	0.014422854	53
917101	0.075763889	2382.882986	0.001113018	0.0001298521	1	0.165236748	1	0.769327088	0.745443584	0.279923944	0.005425961	52
917122	0.068229167	1151.926505	0.006863609	0.032834021	1	0.070027362	1	0.769327088	0.745443584	0.11213653	0.014422854	61
917125	0.008055556	1152.15728	0.006863609	0.032834021	1	0.070027362	1	0.769327088	0.745443584	0.151138524	0.014561981	67
917208	0.335902778	3416.112662	0.0009277515	0.003199926	1	0.023651625	1	0.769327088	0.745443584	0.151138524	0.014422854	36
917266	0.395347222	1151.041551	0.006863609	0.030607986	1	0.02077633	1	0.769327088	0.745443584	0.11213653	0.014561981	55
917293	0.245810185	2381.869699	0.006863609	0.080044521	1	0.165236748	1	0.769327088	0.745443584	0.151138524	0.005425961	60
917370	0.90837963	2380.928738	0.003988313	0.004869452	1	0.165236748	1	0.156147104	0.22144414	0.135231647	0.014422854	122
917534	0.726828704	3413.836493	0.003988313	0.00816213	1	0.165236748	1	0.156147104	0.22144414	0.135231647	0.014422854	65
917657	0.140081019	1149.014583	0.004591198	0.006678106	1	0.010759171	1	0.156147104	0.22144414	0.279923944	0.014422854	63
917979	0.808726852	2380.22669	0.000185503	0.003199926	1	0.023651625	1	0.769327088	0.745443584	0.11213653	0.014422854	77
918113	0.222094907	2378.973484	0.003988313	0.011083801	1	0.165236748	1	0.156147104	0.22144414	0.279923944	0.014422854	120
918115	0.002511574	604.3510764	4.64E-05	0.080044521	1	0.035060057	1	0.769327088	0.745443584	0.151138524	0.066085424	53
918159	0.065324074	3412.597558	0.0009277515	0.003199926	1	0.023651625	1	0.769327088	0.745443584	0.151138524	0.014422854	38
918185	0.041365741	1148.498808	0.007327366	0.019570561	1	0.035060057	1	0.769327088	0.745443584	0.11213653	0.005425961	37
918304	0.177152778	29.87722222	0.007327366	0.019570561	1	0.035060057	1	0.769327088	0.745443584	0.151138524	0.005425961	78
918504	0.635115741	0.064363426	0.004591198	0.080044521	1	0.035060057	1	0.156147104	0.22144414	0.321569355	0.014422854	51
918530	0.043865741	2378.13265	0.004591198	0.011083801	1	0.165236748	1	0.769327088	0.745443584	0.135231647	0.014422854	72
918721	0.212743056	579.2460417	0.011222928	0.00816213	1	0.016602514	1	0.769327088	0.22144414	0.135231647	0.005425961	51
918761	0.067013889	2377.724664	0.000602885	0.0001113018	1	0.165236748	1	0.769327088	0.745443584	0.321569355	0.014422854	33
919046	0.737152778	2377.003958	0.002550665	0.004452071	1	0.165236748	1	0.769327088	0.745443584	0.279923944	0.014422854	47
919071	0.036377315	28.14523148	0.008672263	0.00250429	1	0.035060057	1	0.769327088	0.22144414	0.321569355	0.005425961	49
919074	0.003564815	2376.943831	0.0003710006	0.00816213	1	0.165236748	1	0.156147104	0.22144414	0.135231647	0.005425961	68
919106	0.040636574	2377.939294	0.006863609	0.004869452	1	0.165236748	1	0.769327088	0.745443584	0.151138524	0.005425961	68
919181	0.160069444	1145.908565	0.004591198	0.000556509	1	0.119000139	1	0.769327088	0.745443584	0.321569355	0.014422854	75

(continued)

Table 8.3 (continued)

Bug ID	Opened	Changed	Reporter	Product	Component	Status	Resolution	Hardware	OS	Severity	Version	Summary
919439	0.74474537	2376.057037	0.006863609	0.004869452	1	0.165236748	1	0.769327088	0.745443584	0.151138524	0.005425961	37
919443	0.013518519	2376.009491	0.006863609	0.080044521	1	0.165236748	1	0.769327088	0.745443584	0.151138524	0.005425961	39
919497	0.118055556	1145.323032	0.0001901405	0.000510133	1	0.02077633	1	0.769327088	0.745443584	0.135231647	0.014422854	50
919526	0.058923611	3409.50919	0.0001901405	0.00250429	1	0.023651625	1	0.769327088	0.745443584	0.321569355	0.014422854	72
919607	0.165289352	286.0890741	0.004544822	0.013402588	1	0.035060057	1	0.769327088	0.745443584	0.151138524	0.014422854	64
919799	1.406435185	2374.398206	0.004591198	0.080044521	1	0.165236748	1	0.769327088	0.745443584	0.135231647	0.014422854	50
919868	0.333078704	2374.055741	0.002550665	0.004452071	1	0.165236748	1	0.769327088	0.745443584	0.279923944	0.014422854	93
920024	0.682013889	2373.371134	0.004591198	0.011083801	1	0.119000139	1	0.769327088	0.745443584	0.11213653	0.014422854	32
920213	0.302418981	24.1175	0.004591198	0.00250429	1	0.035060057	1	0.156147104	0.22144414	0.11213653	0.014422854	53
920230	0.019675926	69.74320602	0.004591198	0.019570561	1	0.035060057	1	0.156147104	0.22144414	0.321569355	0.014422854	36
920282	0.062083333	2372.91838	0.003988313	0.004869452	1	0.165236748	1	0.156147104	0.22144414	0.321569355	0.014422854	112
920291	0.020671296	78.89256944	0.004452071	0.080044521	1	0.035060057	1	0.769327088	0.745443584	0.11213653	0.014422854	38
920312	0.061388889	1142.311053	0.007373742	0.032834021	1	0.070027362	1	0.769327088	0.745443584	0.279923944	0.014422854	50
920328	0.054814815	2372.785822	4.64E-05	0.011083801	1	0.165236748	1	0.769327088	0.745443584	0.11213653	0.014422854	80
920366	0.059236111	2372.677569	0.007373742	0.032834021	1	0.070027362	1	0.769327088	0.745443584	0.151138524	0.005425961	68
920516	0.500474537	91.37215278	0.008672263	0.019570561	1	0.035060057	1	0.769327088	0.22144414	0.279923944	0.005425961	78
920638	0.148958333	1141.213553	0.000278254	0.000556509	1	0.119000139	1	0.769327088	0.745443584	0.321569355	0.014422854	51
920704	0.07255787	2371.929676	0.002457914	0.019570561	1	0.035060057	1	0.769327088	0.745443584	0.321569355	0.014422854	40
920738	0.034097222	152.0278819	0.007373742	0.080044521	1	0.016602514	1	0.769327088	0.745443584	0.151138524	0.005425961	41
920867	0.263541667	1140.947685	0.004452071	0.00250429	1	0.035060057	1	0.769327088	0.745443584	0.11213653	0.014422854	75
920982	0.439641204	1140.558171	0.001484024	0.00881139	1	0.02077633	1	0.769327088	0.745443584	0.151138524	0.0330659	41
921517	1.093668981	2370.081493	0.002643417	0.004869452	1	0.165236748	1	0.769327088	0.745443584	0.11213653	0.005425961	29
921692	0.248090278	0.822662037	0.001066642	0.080044521	1	0.035060057	1	0.156147104	0.745443584	0.151138524	0.014422854	48
923086	4.603391204	2365.292523	0.004591198	0.004869452	1	0.165236748	1	0.769327088	0.745443584	0.11213653	0.005425961	77
923115	0.059930556	2365.228229	0.000788388	0.001113018	1	0.165236748	1	0.769327088	0.745443584	0.321569355	0.005425961	60

(continued)

8.4 Numerical Examples

Table 8.3 (continued)

Bug ID	Opened	Changed	Reporter	Product	Component	Status	Resolution	Hardware	OS	Severity	Version	Summary
923226	0.182638889	1134.032164	0.003663683	0.00881139	1	0.02077633	1	0.769327088	0.745443584	0.151138524	0.005425961	54
923395	0.202025463	3398.492442	0.006863609	0.030607986	1	0.02077633	1	0.769327088	0.745443584	0.151138524	0.014422854	69
923426	0.049826389	15.94162037	0.0011222928	0.00816213	1	0.016602514	1	0.769327088	0.745443584	0.279923944	0.005425961	95
923612	0.527708333	1133.827465	0.004591198	0.003199926	1	0.023651625	1	0.769327088	0.745443584	0.279923944	0.005425961	61
923621	0.014328704	2364.233218	0.000185503	0.004869452	1	0.165236748	1	0.156147104	0.22144414	0.321569355	0.014422854	62
923650	0.052916667	2365.184317	0.003988313	0.00250429	1	0.035060057	1	0.156147104	0.22144414	0.279923944	0.014422854	85
923666	0.010092593	1133.634572	0.001484024	0.00881139	1	0.02077633	1	0.769327088	0.745443584	0.279923944	0.014422854	72
923912	0.307928241	1133.0889	0.002133284	0.00250429	1	0.023651625	1	0.769327088	0.745443584	0.11213653	0.005425961	50
924000	0.194224537	2363.7475	0.000371006	0.004452071	1	0.165236748	1	0.769327088	0.745443584	0.321569355	0.005425961	58
924159	0.506215278	14.32834491	0.0013911272	0.00250429	1	0.035060057	1	0.156147104	0.22144414	0.279923944	0.005425961	71
924336	0.203159722	1131.98544	0.008672263	0.019570561	1	0.035060057	1	0.769327088	0.22144414	0.321569355	0.005425961	59
924358	0.025925926	14.09930556	0.008672263	0.00250429	1	0.035060057	1	0.769327088	0.22144414	0.279923944	0.005425961	106
924463	0.190972222	1131.926574	0.000278254	0.006678106	1	0.010759171	1	0.769327088	0.745443584	0.279923944	0.005425961	26
924904	0.957847222	1131.126759	0.000602885	0.080044521	1	0.02077633	1	0.769327088	0.745443584	0.321569355	0.014422854	33
926921	1.475381944	2360.304653	0.000788388	0.004869452	1	0.023651625	1	0.769327088	0.745443584	0.151138524	0.005425961	48
927167	1.111226852	2359.325706	0.00282892	0.080044521	1	0.165236748	1	0.769327088	0.22144414	0.279923944	0.014422854	100
927349	0.350474537	2358.889803	0.01479386	0.011083801	1	0.165236748	1	0.156147104	0.22144414	0.321569355	0.066085424	53
927381	0.078784722	1127.905764	0.0011222928	0.00881139	1	0.02077633	1	0.769327088	0.745443584	0.279923944	0.005425961	59
927423	0.066180556	1127.854931	0.0011222928	0.006678106	1	0.010759171	1	0.769327088	0.745443584	0.279923944	0.005425961	50
927929	0.713414352	1127.001424	0.000927515	0.003199926	1	0.023651625	1	0.769327088	0.745443584	0.321569355	0.005425961	42
928040	0.183946759	1126.786273	0.000463757	0.005054955	1	0.035060057	1	0.769327088	0.745443584	0.151138524	0.014422854	91
928424	0.881967593	1126.430683	0.000463757	0.005054955	1	0.035060057	1	0.769327088	0.745443584	0.151138524	0.014422854	50
928491	0.109421296	33.7383912	0.001576775	0.080044521	1	0.035060057	1	0.769327088	0.745443584	0.151138524	0.005425961	146
928757	0.719027778	2356.22794	4.64E-05	0.011083801	1	0.165236748	1	0.769327088	0.745443584	0.151138524	0.014422854	51
928822	0.098576389	62.03891204	0.004544822	0.080044521	1	0.035060057	1	0.769327088	0.745443584	0.279923944	0.014422854	83

(continued)

Table 8.3 (continued)

Bug ID	Opened	Changed	Reporter	Product	Component	Status	Resolution	Hardware	OS	Severity	Version	Summary
928969	0.293425926	1124.733333	0.004591198	0.019570561	1	0.035060057	1	0.156147104	0.22144414	0.279923944	0.014422854	56
928974	0.029178241	0.02875	0.004591198	0.080044521	1	0.035060057	1	0.156147104	0.745443584	0.279923944	0.014422854	76
929089	0.453981481	74.41753472	0.004591198	0.019570561	1	0.035060057	1	0.156147104	0.22144414	0.279923944	0.014422854	29
929108	0.025972222	1124.258935	0.001484024	0.080044521	1	0.02077633	1	0.769327088	0.745443584	0.151138524	0.014561981	56
929194	0.153773148	2693.791204	0.001066642	0.004869452	1	0.165236748	1	0.769327088	0.745443584	0.135231647	0.005425961	35
947075	3.048923611	2352.033808	0.00282892	0.011083801	1	0.165236748	1	0.769327088	0.745443584	0.151138524	0.005425961	90
947195	0.218923611	2069.851921	9.28E-05	0.032834021	1	0.070027362	1	0.769327088	0.745443584	0.11213653	0.014422854	47
947351	0.524594907	2351.273576	0.004591198	0.011083801	1	0.165236748	1	0.769327088	0.745443584	0.279923944	0.005425961	66
947381	0.072662037	261.5876736	0.000371006	0.000139127	1	0.035060057	1	0.769327088	0.745443584	0.151138524	0.014561981	59
947624	0.478344907	831.3006019	9.28E-05	0.00250429	1	0.035060057	1	0.769327088	0.22144414	0.279923944	0.005425961	58
948938	2.717974537	2348.071076	4.64E-05	0.004869452	1	0.165236748	1	0.769327088	0.745443584	0.321569355	0.005425961	36
949276	1.86474537	1002.14044	0.000185503	0.003292677	1	0.001947781	1	0.769327088	0.22144414	0.135231647	0.014422854	104
949549	1.065034722	1.107696759	0.003988313	0.019570561	1	0.035060057	1	0.156147104	0.22144414	0.321569355	0.014422854	120
950201	1.30556713	1112.716053	0.000278254	0.009321523	1	0.119000139	1	0.769327088	0.745443584	0.279923944	0.005425961	93
950323	0.378229167	2343.410023	4.64E-05	0.004869452	1	0.165236748	1	0.156147104	0.22144414	0.321569355	0.014422854	63
950886	1.083194444	1111.301597	0.001484024	0.00881139	1	0.070027362	1	0.769327088	0.745443584	0.11213653	0.014561981	52
950890	0.004039352	2342.329815	0.001484024	0.080044521	1	0.165236748	1	0.769327088	0.745443584	0.151138524	0.014561981	52
950891	0.000520833	2342.335475	0.001484024	0.080044521	1	0.165236748	1	0.769327088	0.745443584	0.151138524	0.014561981	52
950975	0.097164352	61.37803241	0.002550665	0.019570561	1	0.035060057	1	0.769327088	0.745443584	0.151138524	0.014561981	51
951505	1.056527778	2341.220532	0.001066642	0.080044521	1	0.165236748	1	0.769327088	0.745443584	0.135231647	0.005425961	46
951510	0.002881944	2341.112338	0.001066642	0.080044521	1	0.165236748	1	0.769327088	0.745443584	0.321569355	0.005425961	46
953136	5.074699074	2336.009769	0.008672263	0.080044521	1	0.165236748	1	0.769327088	0.22144414	0.279923944	0.014561981	59
954172	4.072511574	1101.562975	0.008672263	0.080044521	1	0.001947781	1	0.769327088	0.22144414	0.11213653	0.014561981	100
955155	0.950844907	1100.162882	0.003988313	4.64E-05	1	0.01660251	1	0.156147104	0.22144414	0.321569355	0.014422854	160
955163	0.004664352	3.1253125	0.002457914	0.080044521	1	0.035060057	1	0.769327088	0.745443584	0.279923944	0.005425961	50

(continued)

8.4 Numerical Examples

Table 8.3 (continued)

Bug ID	Opened	Changed	Reporter	Product	Component	Status	Resolution	Hardware	OS	Severity	Version	Summary
955247	0.098240741	1100.331053	0.00064926	0.080044521	1	0.010759171	1	0.769327088	0.745443584	0.151138524	0.014422854	64
956596	2.735381944	2328.291991	0.002550665	0.00064926	1	0.165236748	1	0.769327088	0.745443584	0.11213653	0.014561981	55
956597	0.003136574	1097.731933	0.002550665	0.00250429	1	0.023651625	1	0.769327088	0.745443584	0.321569355	0.014561981	59
956919	0.652453704	1097.007083	9.28E-05	0.032834021	1	0.070027362	1	0.769327088	0.745443584	0.279923944	0.066085424	45
957225	0.638819444	1096.036794	0.001576775	0.019570561	1	0.050060057	1	0.769327088	0.745443584	0.151138524	0.005425961	63
957248	0.066597222	2326.950139	0.006863609	0.011083801	1	0.165236748	1	0.769327088	0.745443584	0.151138524	0.014422854	29
957270	0.05193287	2326.766204	0.002457914	0.011083801	1	0.165236748	1	0.073134536	0.03274127	0.321569355	0.014422854	90
957276	0.013009259	1096.280405	0.000185503	0.0816213	1	0.016602514	1	0.156147104	0.22144414	0.321569355	0.005425961	57
957278	0.004224537	2326.821748	0.002457914	0.004869452	1	0.165236748	1	0.073134536	0.03274127	0.321569355	0.014422854	124
957280	0.003136574	2326.84059	0.002457914	0.080044521	1	0.165236748	1	0.769327088	0.745443584	0.279923944	0.014422854	35
957281	0.006203704	2326.81603	0.002457914	0.011083801	1	0.165236748	1	0.769327088	0.745443584	0.11213653	0.014422854	82
957761	2.719293981	43.22384259	0.002550665	0.019570561	1	0.050060057	1	0.769327088	0.745443584	0.151138524	0.014561981	29
957888	0.267708333	2323.831481	0.002457914	0.080044521	1	0.165236748	1	0.769327088	0.745443584	0.279923944	0.014561981	99
958057	0.57181713	2323.187581	0.001066642	0.001298521	1	0.165236748	1	0.156147104	0.22144414	0.279923944	0.014561981	67
958323	0.528506944	19.90322917	0.001576775	0.080044521	1	0.050060057	1	0.769327088	0.745443584	0.151138524	0.005425961	59
958568	0.92505787	1091.000069	0.000231879	0.009321523	1	0.119000139	1	0.769327088	0.745443584	0.279923944	0.014561981	87
958725	0.564965278	1090.208229	0.002550665	0.00881139	1	0.070027362	1	0.769327088	0.745443584	0.151138524	0.014561981	48
958782	0.079826389	1090.337847	0.002550665	0.00881139	1	0.02077633	1	0.769327088	0.745443584	0.321569355	0.014561981	48
958886	0.139953704	1090.027975	0.007327366	0.009182396	1	0.050060057	1	0.769327088	0.745443584	0.11213653	0.014561981	45
958900	0.019664352	75.0571875	0.007651996	0.019570561	1	0.050060057	1	0.769327088	0.745443584	0.279923944	0.005425961	84
958952	0.122418981	2320.838229	0.004591198	0.004452071	1	0.165236748	1	0.156147104	0.22144414	0.279923944	0.014422854	68
958973	0.063553241	2320.713611	0.002550665	0.004869452	1	0.165236748	1	0.769327088	0.745443584	0.11213653	0.005425961	101
959549	0.933391204	1089.115208	4.64E-05	0.080044521	1	0.023651625	1	0.769327088	0.745443584	0.151138524	0.005425961	63
959736	1.53537037	2318.307072	4.64E-05	0.011083801	1	0.165236748	1	0.769327088	0.745443584	0.151138524	0.014561981	15
959818	0.758657407	2317.575602	0.006863609	0.001113018	1	0.165236748	1	0.769327088	0.745443584	0.151138524	0.005425961	45

(continued)

Figure 8.2 shows the estimated error between validation and training. Also, Figs. 8.3, 8.4, 8.5 and 8.6 are the estimated instantaneous detection time of software faults, the scatter plot of the estimated instantaneous detection time of software faults, the estimated cumulative detection time of software fault, and the scatter plot of the cumulative detection time of software faults, respectively.

8.5 Concluding Remarks

In many open source projects, the bug tracking system such as Bugzilla are helpful for the software developers, managers and users. However, the users as well as the software developer and manager can easily register the fault data by clicking on the dialog screen. Therefore, it is difficult to quickly judge the OSS reliability from the data. Then, it will be able to take quickly action for the software managers in the debugging phase, if the software developers can quickly judge the reliability from the observed fault.

The recognition method of software fault level is discussed in the past. The method of big fault data analysis by using deep learning have been proposed [4]. Especially, it will be difficult for the software developers to quickly judge the fault type by only the data obtained from bug tracking system, because several general software users as well as the main open source project members can report the fault contents to the

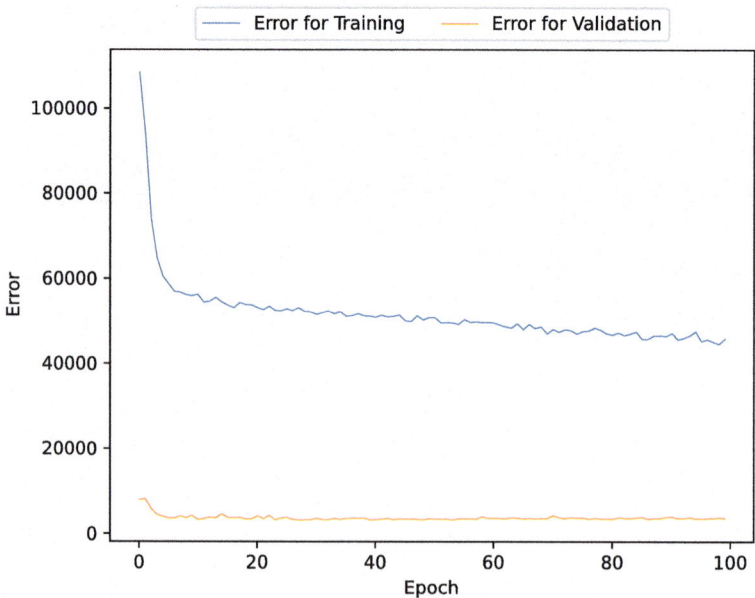

Fig. 8.2 The estimated error between validation and training

8.5 Concluding Remarks

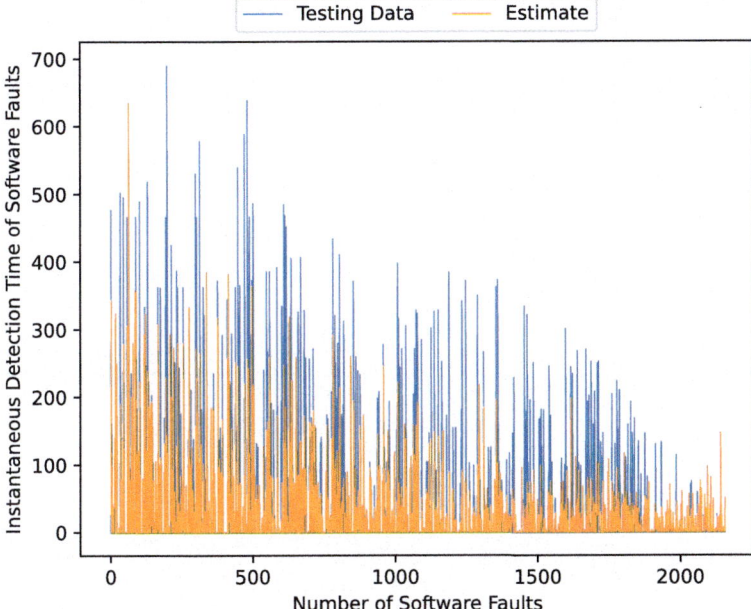

Fig. 8.3 The estimated instantaneous detection time of software faults

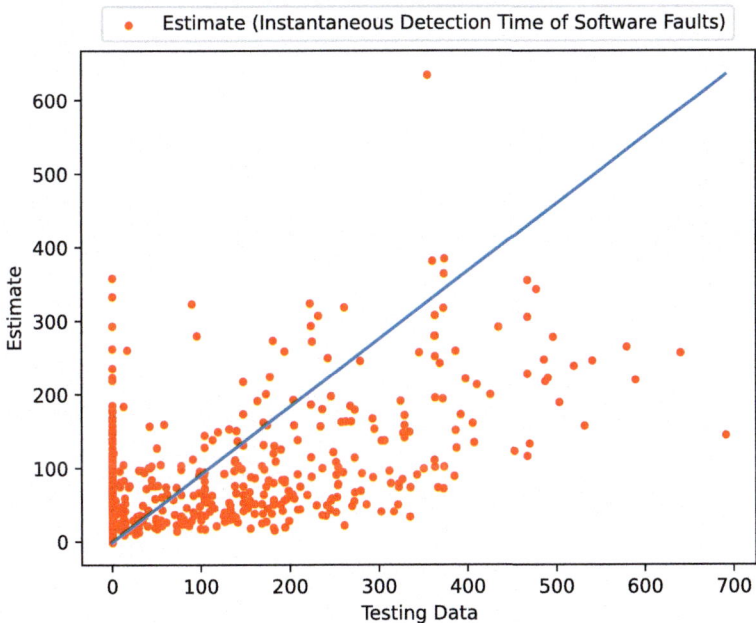

Fig. 8.4 The scatter plot of the estimated instantaneous detection time of software faults

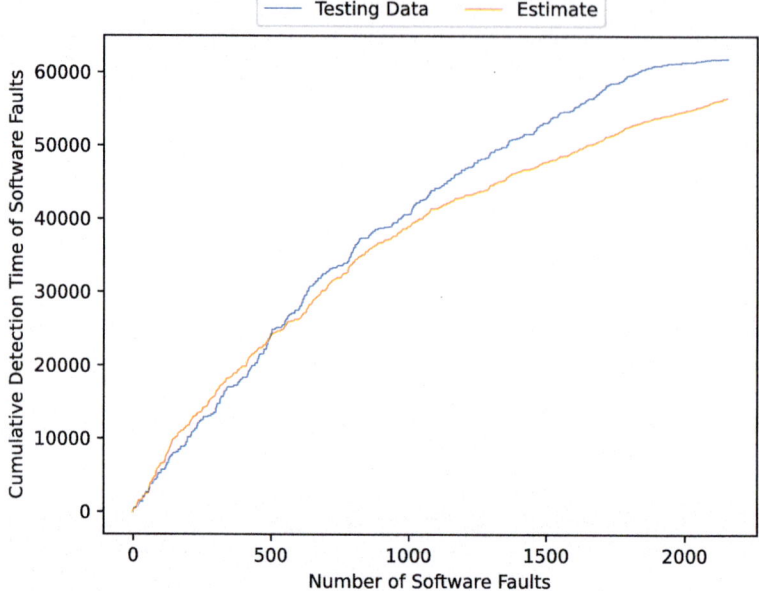

Fig. 8.5 The estimated cumulative detection time of software faults

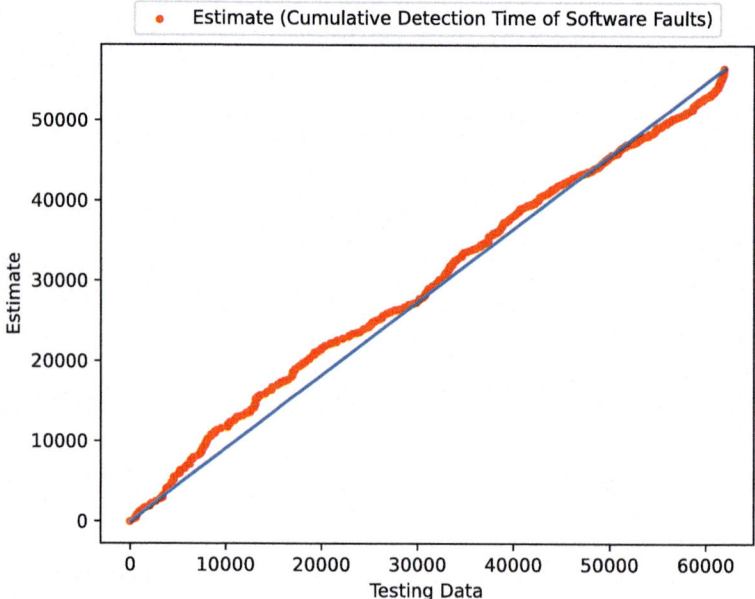

Fig. 8.6 The scatter plot of the cumulative detection time of software faults

bug tracking system. Then, the recognition method of software fault severity levels by using deep learning based on big fault data have been introduced in this chapter [4].

As the other approach, we have proposed the method of reliability assessment as the case study of the detection time of software faults for the objective variable. Then, the input data sets are 13 kinds of factors from the bug tracking system. Moreover, several performance examples of the proposed method by using the big fault data in the actual open source software project have been shown in this chapter.

References

1. Yamada S (2014) Software reliability modeling: fundamentals and ApplicatIONS. Springer, Tokyo/Heidelberg
2. Kapur PK, Pham H, Gupta A, Jha PC (2011) Software reliability assessment with OR applications. Springer, London
3. Yamada S, Tamura Y (2016) OSS reliability measurement and assessment. Springer series in reliability engineering. Springer
4. Tamura Y, Ashida S, Matsumoto M, Yamada S (2016) Identification method of fault level based on deep learning for open source software. In: Software engineering research, management and applications, Studies in computational intelligence. Springer International Publishing Switzerland, pp 65–76
5. Karnin ED (1990) A simple procedure for pruning back-propagation trained neural networks. IEEE Trans Neural Netw 1:239–242
6. Kingma DP, Rezende DJ, Mohamed S, Welling M (2014) Semi-supervised learning with deep generative models. In: Proceedings of neural information processing systems, pp 3581–3589
7. Blum A, Lafferty J, Rwebangira MR, Reddy R (2004) Semi-supervised learning using randomized mincuts. In: Proceedings of the international conference on machine learning, pp 1–13
8. George ED, Dong Y, Li D, Alex A (2012) Context-dependent pre-trained deep neural networks for large-vocabulary speech recognition. IEEE Trans Audio Speech Lang Process 20(1):30–42
9. Vincent P, Larochelle H, Lajoie I, Bengio Y, Manzagol PA (2010) Stacked denoising autoencoders: learning useful representations in a deep network with a local denoising criterion. J Mach Learn Res 11(2):3371–3408
10. Martinez HP, Bengio Y, Yannakakis GN (2013) Learning deep physiological models of affect. IEEE Comput Intell Mag 8(2):20–33
11. Hutchinson B, Deng L, Yu D (2013) Tensor deep stacking networks. IEEE Trans Pattern Anal Mach Intell 35(8):1944–1957
12. Tamura Y, Ueki R, Anand A, Yamada S (2023) Estimation and comparison of mean time between failures based on deep learning for OSS fault big data. Int J Syst Assur Eng Manage. Springer, Latest Articles, https://doi.org/10.1007/s13198-023-01907-2, pp 1–16, April 2023
13. Tamura Y, Yamada S (2022) Prototype of 3D reliability assessment tool based on deep learning for edge OSS computing. Mathematics, vol 10, No 9, Multidisciplinary Digital Publishing Institute, Switzerland, https://doi.org/10.3390/math10091572, pp 1–12, May 2022
14. The OpenStack project, Build the future of Open Infrastructure, https://www.openstack.org/

Deep Learning Approach for OSS Reliability Assessment Considering Wiener Process

9.1 Introduction

The OSS have been developed under the initiative of many corporate organizations. Also, many OSS's have been maintained by many corporate organizations. Especially, many OSS have been developed by using the bug tracking systems. The specified bug tracking systems have been used by several OSS projects. Also, the fault big data sets recorded on the bug tracking system will be very useful to assess the reliability of OSS, because the cumulative number of detected software faults is only used in order to assess the typical software reliability in the past. On the other hand, we can use various fault big data sets obtained from the bug tracking system in case of OSS system.

OSS have been used in various areas of social systems. In particular, many OSS's are used in various areas, e.g., Apache HTTP server, Apache Tomcat, OpenStack, Java, Ruby, WordPress, and PHP, etc. Recently, the cloud computing service has been mainly used as the network service [1]. The network service is changing from the cloud computing to the edge computing [2–6]. The edge computing is useful for the continuity and responsibility of network service. Thus, OSS such as OpenStack is used in the area of edge computing.

In this chapter, we focus on the OSS reliability assessment based on the OSS fault under the edge computing. In particular, we cover the fault modification in terms of effort management. Also, several research papers in terms of the reliability of OSS have been proposed by several researchers [7]. In the past, the software reliability has been generally assessed by using the SRGM's. Then, the methods of software reliability assessment based on the SRGM's have been proposed by several researchers [8–11]. On the other hand, it is important to appropriately control the operation in the case of the edge computing. The optimal control of edge OSS computing will indirectly relate to the reliability and cost optimization.

© The Author(s), under exclusive license to Springer Nature Switzerland AG 2024
Y. Tamura and S. Yamada, *Applied OSS Reliability Assessment Modeling, AI and Tools*, Springer Series in Reliability Engineering,
https://doi.org/10.1007/978-3-031-64803-8_9

This chapter proposes the OSS reliability assessment based on the deep learning considering the fault modification in the edge OSS operation. We consider that the edge OSS operation depends on various factors of edge computing. Moreover, we show several numerical examples based on the proposed deep learning model considering the environment of imperfect debugging.

9.2 Wiener Deep Learning Approach

We consider the optimal fault modification policy based on the deep learning. Then, we apply the fault-detection time to the objective variable of deep learning. The OSS fault data are used for the software reliability by using many SRGM's.

This chapter focuses on the fault-detection time. We consider that the proposed deep learning estimates the change of the situation of fault-detection by learning the phenomena of fault-detection.

Then, we consider the following fault-detection time:

$$O^h(i) \Leftarrow F^h(i), \tag{9.1}$$
$$\text{subject to} \quad F^h(i) \subseteq F(i),$$

where $O^h(i)$ is the OSS fault-detection time in i-th fault in terms of the high fault level, i.e., $O^h(i)$ is the input value and the objective variable. $F(i)$ is all OSS fault-detection time in i-th fault recorded from the bug tracking system. Then, $F(i)$ is included all levels in terms of fault severity. $F^h(i)$ means the OSS fault-detection time in i-th fault in terms of the high fault level.

We focus on the fault big data recorded on the bug tracking system. The fault-detection process is well known as the uncertainty event. For example, there are many SRGM's. Therefore, the fault-detection process is widely assessed by using the SRGM. In this case, the imperfect debugging process is well known as one of the fault-detection process.

Considering $W(i)$ is a Wiener process which is formally defined as an integration of the white noise $\sigma(i)$ with respect to the i-th fault. The Wiener process is a Gaussian process and it has the following properties:

$$\Pr[W(0) = 0] = 1, \tag{9.2}$$
$$E[W(i)] = 0, \tag{9.3}$$
$$E[W(i)W(i')] = \text{Min}[i, i'], \tag{9.4}$$

where $\Pr[\cdot]$ and $E[\cdot]$ represent the probability and expectation, respectively.

Generally, the Wiener process is considered as the continuous time. On the other hand, this chapter focuses the discrete phenomenon for the fault-detection process. Then, there is no problem that the SRGM's based on Wiener process have been proposed as the unit time. We assume that the fault-detection process is approximately based on Wiener process. We consider the following equation as the objective variable.

$$O_w(i) = O^h(i) + W(i), \tag{9.5}$$

where $W(i)$ is the Wiener process at the i-th fault. This equation is applied to the data preprocessing and the output values of the training data and testing one.

9.3 Numerical Examples

We focus on *OpenStack* Project [12] including several edge components. Table 8.2 is a part of data set. All lines of data is over 20,000 lines. Then, the all data is over 260,000 data items. Actually, the users can obtain a greater number of data according to various OSS projects. In this chapter, we show numerical examples by using a part of data set as shown in Table 8.3 on the assumption of the edge OSS service.

Figure 9.1 is the model structure of the proposed deep learning. Figure 9.1 means that the number of input unit is 13, the number of objective unit is 1, the number of layer is 5, and the unit for each layer is 100, respectively. Considering the case with Wiener process, Figs. 9.2, 9.3, 9.4, 9.5, and 9.6 show the estimated error for validation and training data, the estimated frequency of fault-detection time, the estimated scatter plot of fault-detection time, the estimated cumulative fault-detection time, and the estimated cumulative scatter plot for fault-detection time.

Similarly, considering the case without Wiener process, Figs. 9.7, 9.8, 9.9, 9.10, and 9.11 show the estimated error for validation and training data, the estimated frequency of fault-detection time, the estimated scatter plot of fault-detection time, the estimated cumulative fault-detection time, and the estimated cumulative scatter plot for fault-detection time.

From above mentioned results, we have found that our deep learning model considering the Wiener process can describe the characteristics of edge computing according to the changes of the noisy cases in the fault-detection. In particular, the noisy approach will be useful to estimate the uncertainty event such as the imperfect debugging.

We show the estimation results of sensitivity analysis. Figures 9.12, 9.13, 9.14, and 9.15 show the sensitivity analysis of the Wiener term in the case of Wiener process, the sensitivity analysis of the Wiener term in the case of 2.5%, the sensitivity analysis of the Wiener term in the case of 5.0%, and the sensitivity analysis of the Wiener term in the case of 10%, respectively.

Fig. 9.1 The model structure of the proposed deep learning

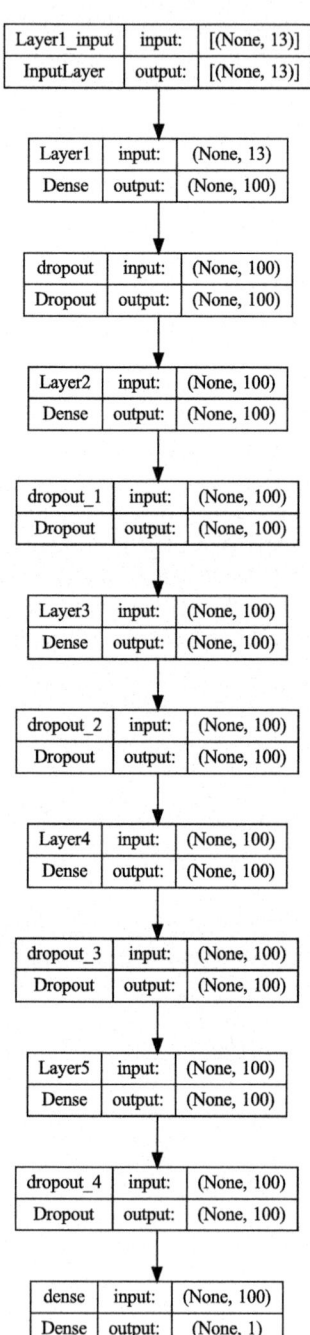

9.3 Numerical Examples

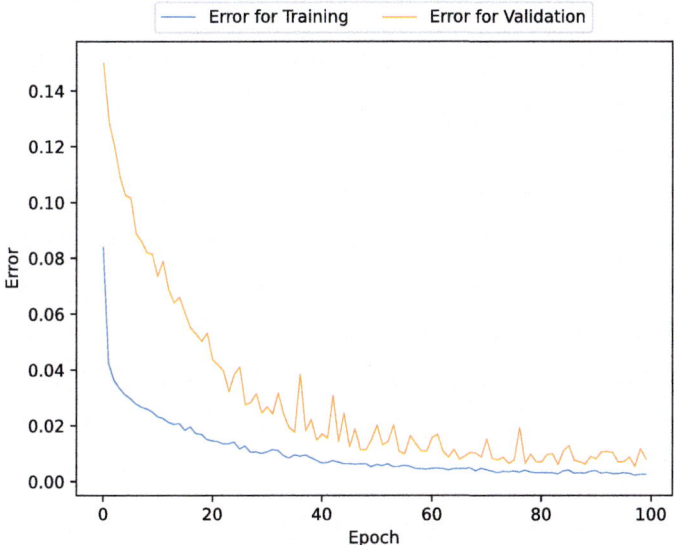

Fig. 9.2 The estimated error for validation and training data in the case of Wiener process

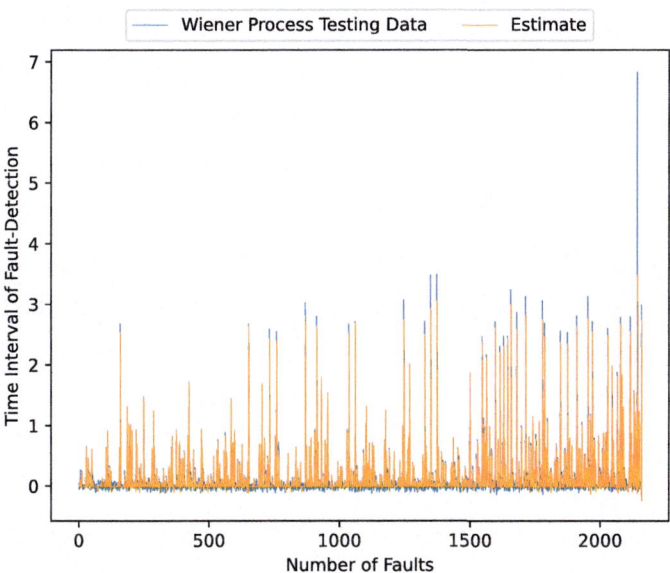

Fig. 9.3 The estimated frequency of fault-detection times in the case of Wiener process

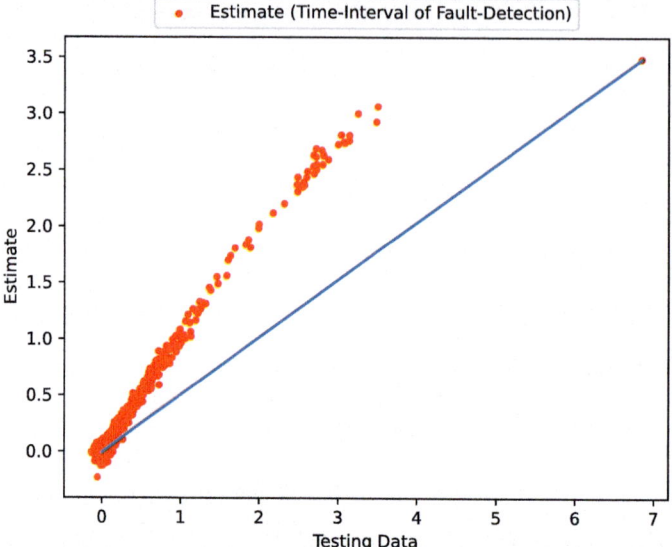

Fig. 9.4 The estimated scatter plot of fault-detection times in the case of Wiener process

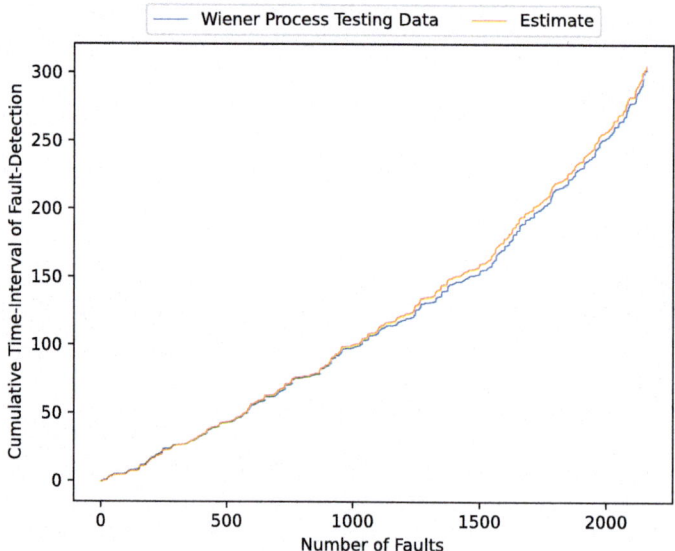

Fig. 9.5 The estimated cumulative fault-detection times in the case of Wiener process

9.3 Numerical Examples

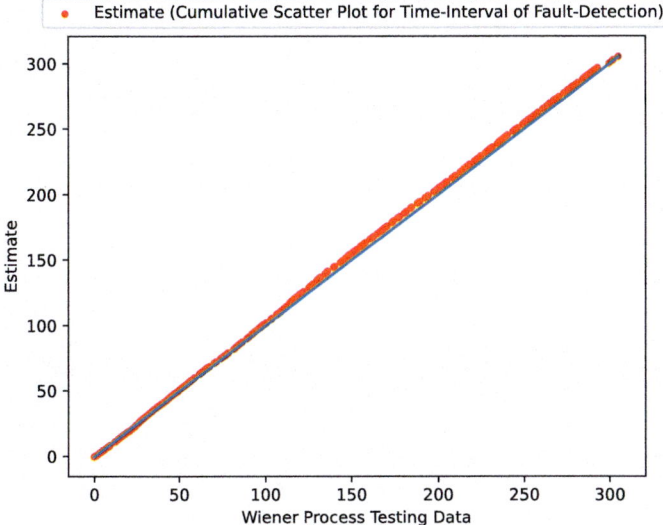

Fig. 9.6 The estimated cumulative scatter plot for fault-detection times in the case of Wiener process

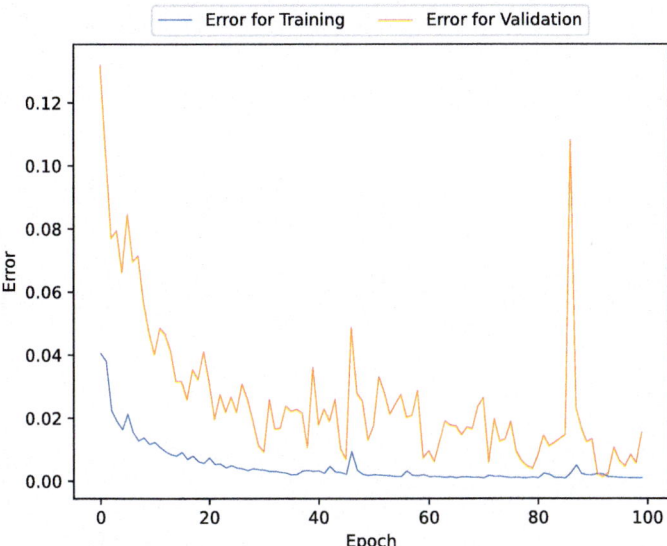

Fig. 9.7 The estimated error for validation and training data without the case of Wiener process

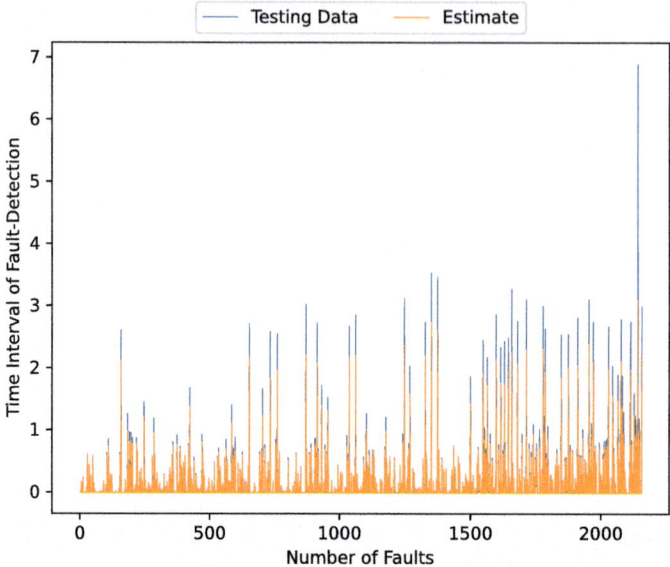

Fig. 9.8 The estimated frequency of fault-detection times without the case of Wiener process

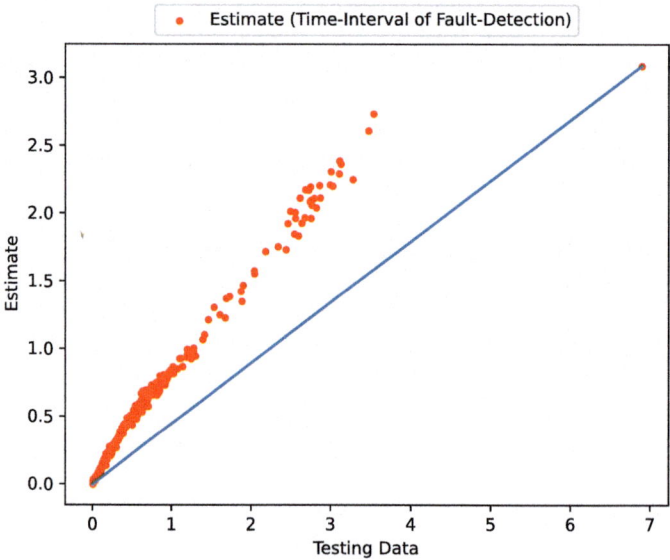

Fig. 9.9 The estimated scatter plot of fault-detection times without the case of Wiener process

9.3 Numerical Examples

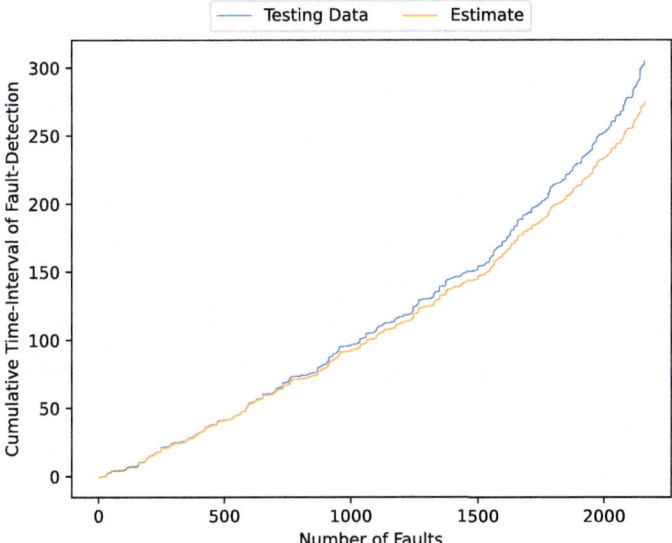

Fig. 9.10 The estimated cumulative fault-detection times without the case of Wiener process

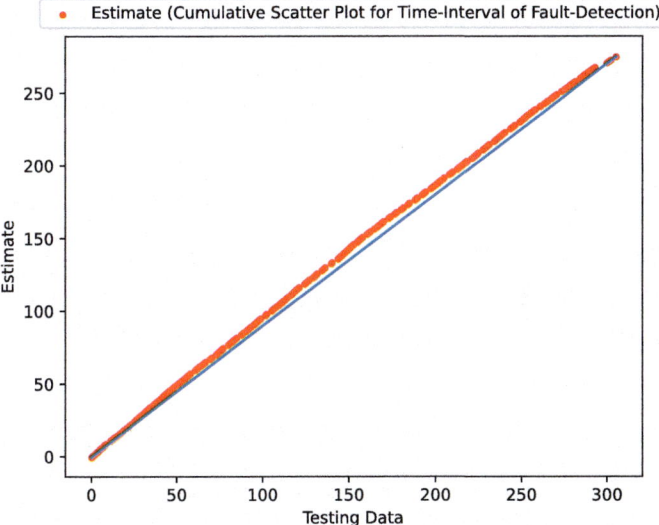

Fig. 9.11 The estimated cumulative scatter plot for fault-detection times without the case of Wiener process

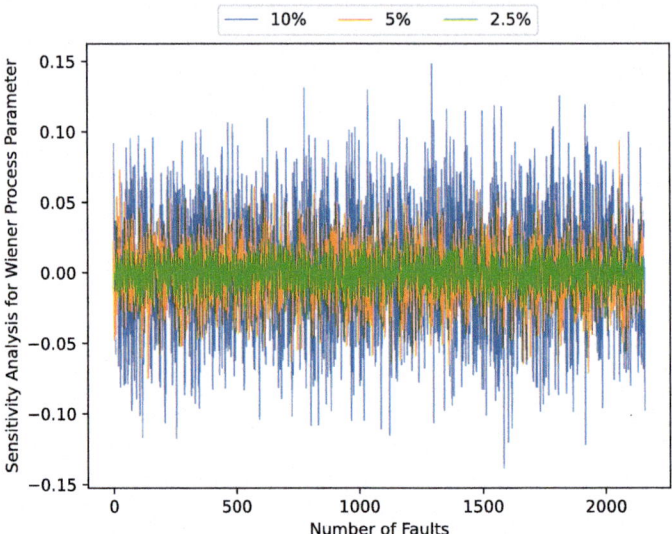

Fig. 9.12 The sensitivity analysis of the Wiener term in the case of Wiener process

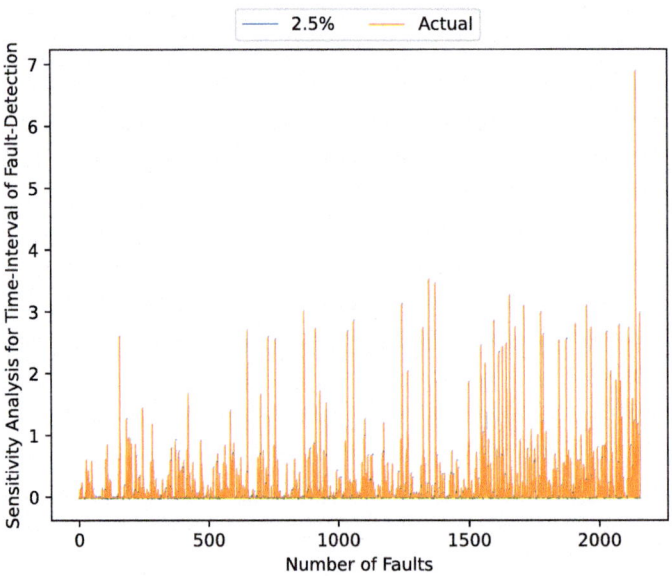

Fig. 9.13 The sensitivity analysis of the Wiener term in the case of 2.5%

9.3 Numerical Examples

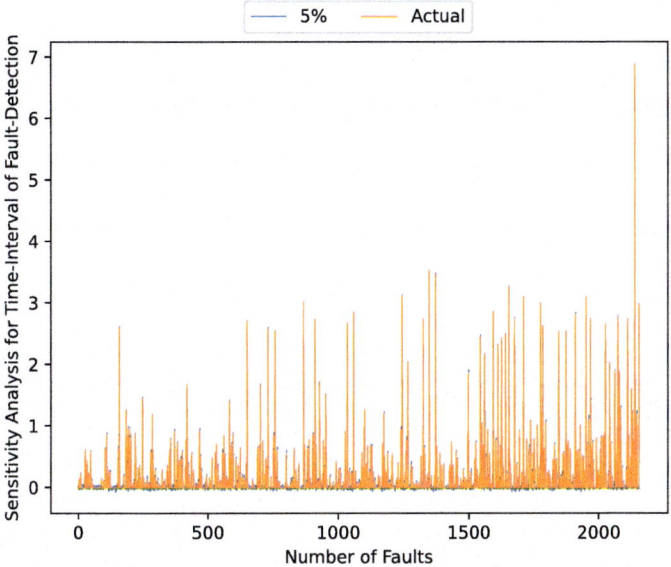

Fig. 9.14 The sensitivity analysis of the Wiener term in the case of 5.0%

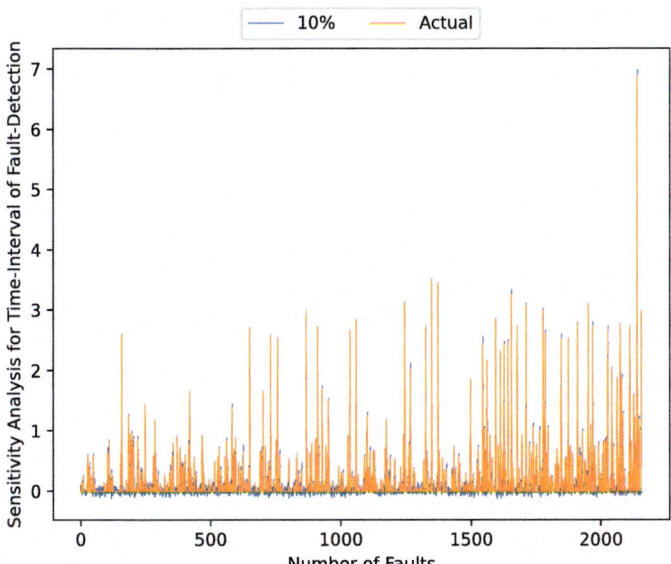

Fig. 9.15 The sensitivity analysis of the Wiener term in the case of 10%

9.4 Concluding Remarks

The appropriate reliability control for the maintenance of edge software service will indirectly depends on the placement environment of edge server. This chapter has proposed the deep learning model considering the irregular fluctuations as the characteristics of edge server and OSS. It is difficult for the edge server managers to control the progress of server maintenance. This chapter has discussed the deep learning based on the characteristics of edge and OSS by using the actual edge OSS fault-detection data as follows:

- ◈ Edge OSS reliability depends on the OSS fault-detection process.
- ◈ The OSS reliability has the noise such as the Wiener processes.
- ◈ The data preprocessing in the fault-detection process differ from the general data in the open data sets.
- ◈ The imperfect debugging process is considered by using the Wiener process.

The proposed model will be useful as the assessment measures of the reliability for edge OSS service in operation phase.

On the other hand, we will consider that the data for frequency of fault-detection takes the negative values because of the Wiener process. This will be needed the consideration for this. Then, we consider that this negative value means the surplus value. Therefore, there is no problem in the case of negative value.

References

1. Ibrahim IM et al (2018) A robust generic multi-authority attributes management system for cloud storage services. IEEE Trans Cloud Comput. https://doi.org/10.1109/TCC.2018.2867871
2. Ahmad AA et al (2019) Scalability analysis comparisons of cloud-based software services. J Cloud Comput Adv Syst Appl. https://doi.org/10.1186/s13677-019-0134-y
3. Ozcan MO, Odaci F, Ari I (2019) Remote debugging for containerized applications in edge computing environments. In: Proceedings of the 2019 IEEE international conference on edge computing (EDGE), Milan, Italy, pp 30–32. https://doi.org/10.1109/EDGE.2019.00021
4. Ngoko Y, Cérin C (2017) An edge computing platform for the detection of acoustic events. In: Proceedings of the 2017 IEEE international conference on edge computing (EDGE), Honolulu, HI, USA, pp 240–243, https://doi.org/10.1109/IEEE.EDGE.2017.44
5. Caprolu M, Di Pietro R, Lombardi F, Raponi S (2019) Edge computing perspectives: architectures, technologies, and open security issues. In: Proceedings of the 2019 IEEE international conference on edge computing (EDGE), Milan, Italy, pp 116–123. https://doi.org/10.1109/EDGE.2019.00035
6. Dolui K, Datta SK (2017) Comparison of edge computing implementations: fog computing, cloudlet and mobile edge computing. In: Proceedings of the 2017 global internet of things summit (GIoTS), Geneva, Switzerland, pp 1–6. https://doi.org/10.1109/GIOTS.2017.8016213
7. Yamada S, Tamura Y (2016) OSS reliability measurement and assessment. Springer series in reliability engineering. Springer

8. Yamada S (2014) Software reliability modeling: fundamentals and applications. Springer, Tokyo/Heidelberg
9. Lyu MR (ed) (1996) Handbook of software reliability engineering. IEEE Computer Society Press, Los Alamitos, CA, U.S.A
10. Musa JD, Iannino A, Okumoto K (1987) Software reliability: measurement, prediction, application. McGraw-Hill, New York
11. Kapur PK, Pham H, Gupta A, Jha PC (2011) Software reliability assessment with OR applications. Springer, London
12. The OpenStack project, Build the future of Open Infrastructure. https://www.openstack.org/

Deep Learning Approach for OSS Reliability Assessment Considering Jump Diffusion Process

10.1 Introduction

In this chapter, we discuss the method of reliability assessment for OSS. In particular, we focus on the deep learning as the method of OSS reliability assessment. Many OSS's have been developed under several OSS development projects. At present, many source codes of OSS have been used for the source codes of many proprietary software. At present, the cloud computing service has been mainly used as the network service [1] by many users. The network service is changing from the cloud computing to the edge computing [2–6]. The edge computing will be useful for the continuity and responsibility of network service. In particular, OSS such as OpenStack has been used in the area of edge computing.

In this chapter, we focus on the reliability assessment for the OSS. In particular, we focus on the reliability in terms of OSS fault-detection time. Also, several research papers in terms of the reliability of OSS have been proposed by several researchers [7]. In the past, the software reliability has been generally assessed by using SRGM's. Then, the methods of software reliability assessment based on the SRGM's have been proposed in several research papers [8–11].

Then, we propose a method of reliability assessment based on the deep learning. Also, we show several numerical examples by using the fault big data sets of OpenStack as OSS. This chapter proposes the OSS reliability assessment based on the deep learning and jump diffusion process for the fault big data in the edge OSS operation. In the past, we have shown several numerical examples based on the proposed deep learning model considering the environment of imperfect debugging by using the Wiener process. On the other hand, we have shown several numerical examples based on the proposed deep learning model considering the environment of imperfect debugging by using the jump diffusion process. Then, we have proposed the SRGM's based on the jump diffusion process [12–17]. In this chapter, we use the deep learning approach based on the jump diffusion process modeling.

10.2 OSS Reliability Assessment Based on Deep Learning Model with Jump Diffusion Process

We consider the fault-detection process based on the deep learning. Then, we apply the fault-detection time to the objective variable of deep learning. The software fault data are used for the software reliability by using many SRGM's. In particular, the fault-detection time data is used for the method of reliability assessment the software reliability assessment models based on the hazard rates. This chapter focuses on the fault-detection time.

Then, we consider the following fault-detection phenomenon.

$$d_i = o_i - o_{i-1}, \tag{10.1}$$

where d_i is the time-interval of fault-detection. o_i ($i = 1, 2, \cdots, n$) is the OSS fault-detection time between i-th and $i - 1$-th fault until the end of n-th fault. This means that the reliability of OSS grows, if d_i becomes large.

Moreover, we consider the jump phenomena as the irregular fluctuation influenced for the OSS fault-detection. For example, the sample path of jump term [13,14,18] is given by

$$N_i = \sum_{i=1}^{M_i(\rho)} \log J_j, \tag{10.2}$$

where N_i means the sample path of time-interval for OSS fault-detection at i ($i = 1, 2, \cdots, n$)-th fault. Also, $M_i(\rho)$ is a Poisson point process with parameter ρ. Moreover, $M_i(\rho)$ and ρ are defined as the number of jumps and ρ the jump ratio, respectively. $M_i(\rho)$ and J_i are mutually independent [13,14,18]. J_i represents the range of i-th jump.

The typical jump processes have been used in the financial research area [18]. It is difficult to directly use the typical financial model for the OSS fault-detection process, because the area of financial engineering is based on the log-normal distribution in terms of the jump term. It is difficult to apply the log-normal distribution to the jump term of OSS fault-detection phenomena, because it is usually assumed that the OSS fault-detection situation have non-biased distribution in case of the reliability engineering. Therefore, the following normal distribution is defined in place of Gaussian process of jump considering the OSS fault-detection phenomenon:

$$J_j \equiv f_j(x) = \frac{1}{\sqrt{2\pi}\tau} \exp\left[-\frac{(x-\mu)^2}{2\tau^2}\right], \tag{10.3}$$

where μ is the mean value, τ is the standard deviation. The range J_i for i-th jump are independently defined as the positive values.

Therefore, we consider the objective variable based on the jump diffusion process as

$$d_i^j = d_i + N_i, \tag{10.4}$$

where d_i^j is the objective variable of deep learning with the jump diffusion process.

10.3 Numerical Examples

As several numerical examples, this chapter uses the fault big data sets obtained from the bug tracking system in *OpenStack* Project [19] with several edge components. Table 8.2 is a part of data set. All lines of data is over 20,000 lines. Then, the all data is over 260,000 data items. Actually, the users can obtain a greater number of data according to various OSS projects. In this chapter, we show numerical examples by using a part of data set as shown in Table 8.3 on the assumption of the edge OSS service.

Figure 10.1 is the model structure of the deep learning used in this chapter. Figure 10.1 means that the number of input unit is 13, the number of objective unit is 1, the number of layer is 5, and the unit for each layer is 100, respectively. Considering the case with jump diffusion process, Figs. 10.2, 10.3, 10.4, 10.5 and 10.6 show the estimated error for validation and training data in the case of jump diffusion process, the estimated time-interval of fault-detection in the case of jump diffusion process, the estimated scatter plot of time-interval of fault-detection in the case of jump diffusion process, the estimated cumulative time-interval of fault-detection in the case of jump diffusion process, and the estimated scatter plot for the cumulative time-interval of fault-detection in the case of jump diffusion process, respectively.

On the other hand, as the case without jump diffusion process, Figs. 10.7, 10.8, 10.9, 10.10 and 10.11 show the estimated error for validation and training data, the estimated time-interval of fault-detection without the case of jump diffusion process, the estimated scatter plot of time-interval of fault-detection without the case of jump diffusion process, the estimated cumulative time-interval of fault-detection without the case of jump diffusion process, and the estimated scatter plot for the cumulative time-interval of fault-detection without the case of jump diffusion process, respectively.

In particular, the sensitivity analysis for the jump term changing the Poisson point process parameter ρ in the case of jump diffusion process is shown in Fig. 10.12. Moreover, this chapter shows the sensitivity analyses for the time-interval of fault-detection changing the Poisson point process parameter ρ in the case of jump diffusion process. The results of sensitivity analyses are shown in Figs. 10.13, 10.14 10.15. From Figs. 10.13, 10.14 and 10.15, we found that the estimation accuracy is good according to the noise based on the jump diffusion process.

From above mentioned results, we have found that the proposed deep learning model considering the jump diffusion process can describe the characteristics of OSS fault-detection phenomenon according to the changes of the noisy cases in the sudden fluctuation. In particular, the noisy approach with jump diffusion process will be useful to assess the OSS reliability with uncertainty event under the OSS fault-detection phenomena.

On the other hand, we will consider that the value of the time-interval of fault-detection takes the negative values because of the jump diffusion process. In terms of this case, the negative value means the imperfect debugging. Therefore, there is no problem in the case of negative value.

Fig. 10.1 The model structure of the proposed deep learning

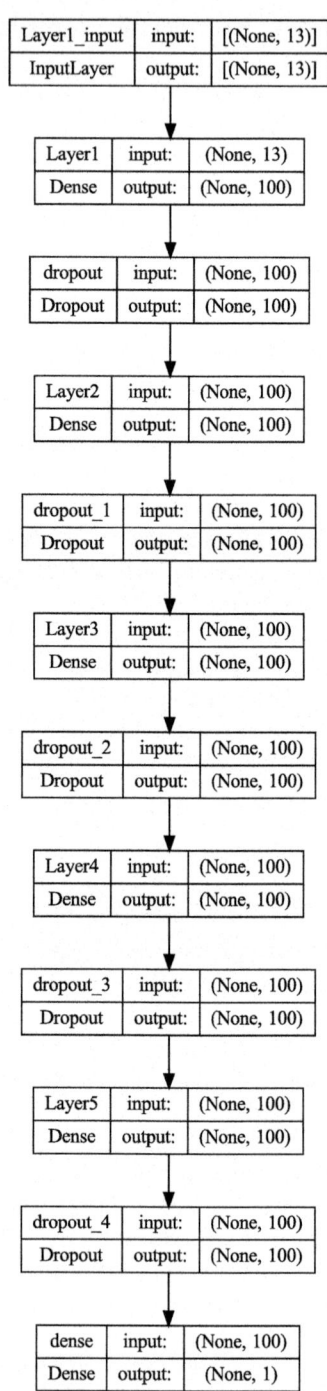

10.3 Numerical Examples

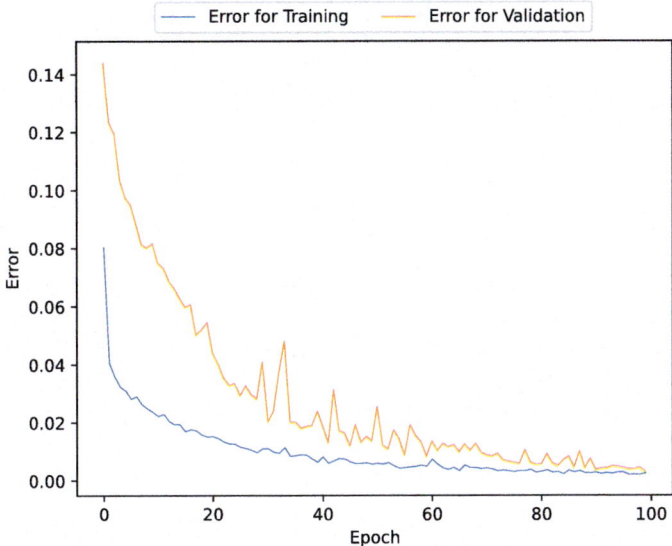

Fig. 10.2 The estimated error for validation and training data in the case of jump diffusion process

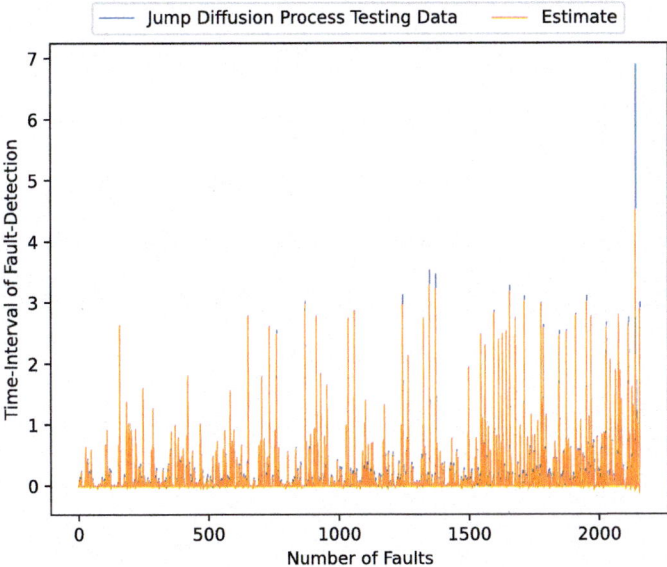

Fig. 10.3 The estimated time-interval of fault-detection in the case of jump diffusion process

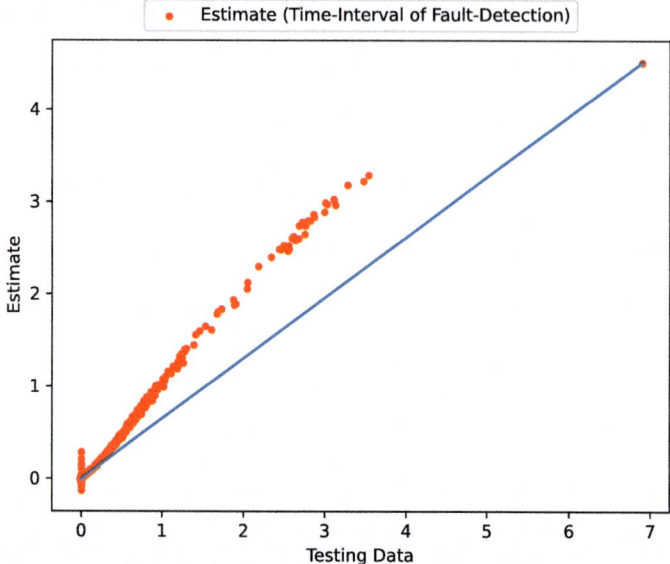

Fig. 10.4 The estimated scatter plot of time-interval of fault-detection in the case of jump diffusion process

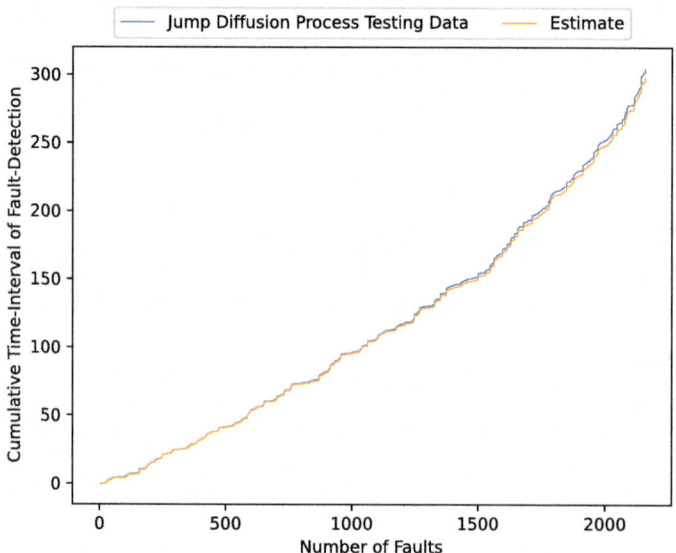

Fig. 10.5 The estimated cumulative time-interval of fault-detection in the case of jump diffusion process

10.3 Numerical Examples

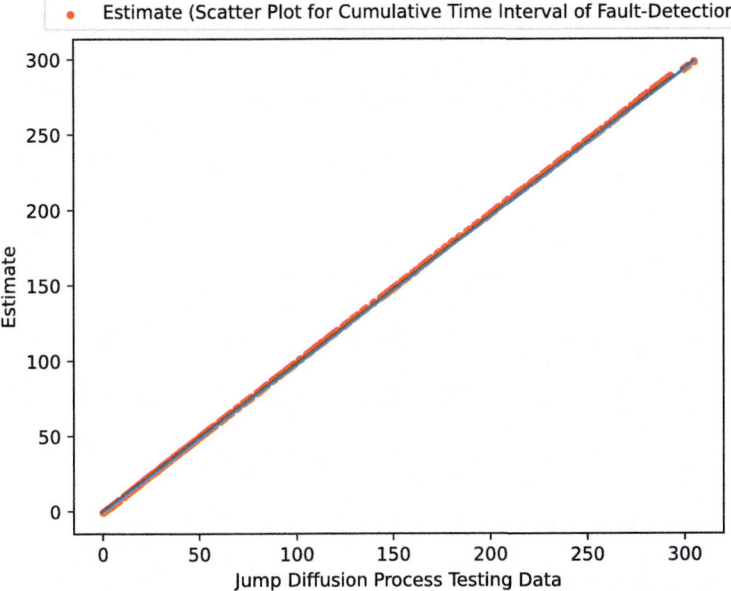

Fig. 10.6 The estimated scatter plot for the cumulative time-interval of fault-detection in the case of jump diffusion process

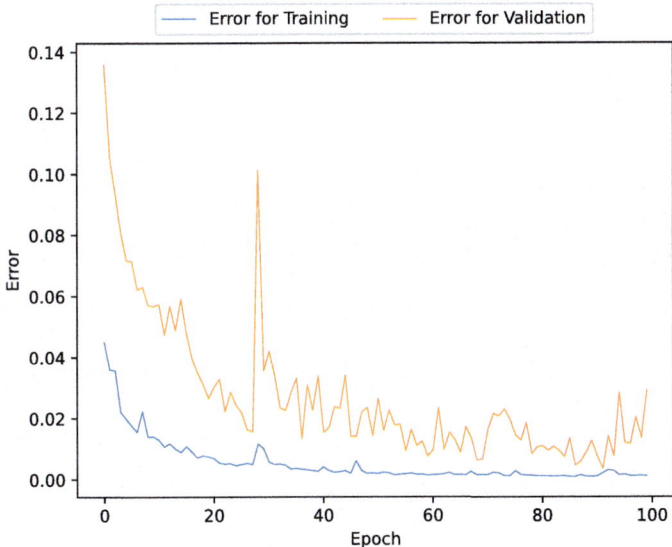

Fig. 10.7 The estimated error for validation and training data without the case of jump diffusion process

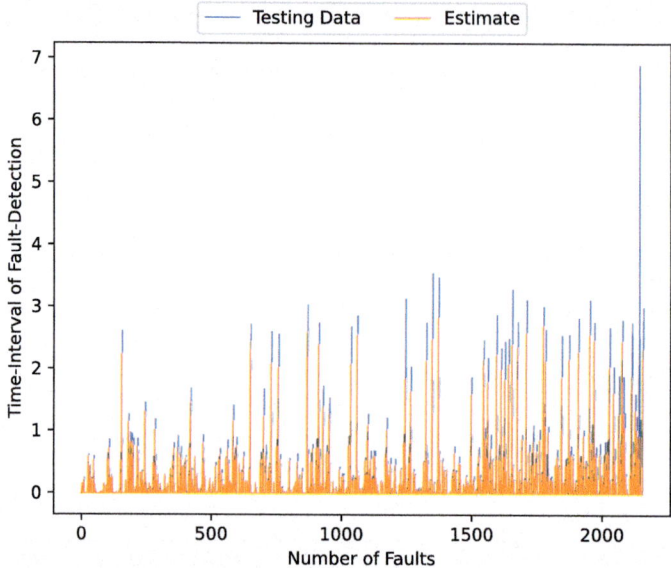

Fig. 10.8 The estimated time-interval of fault-detection without the case of jump diffusion process

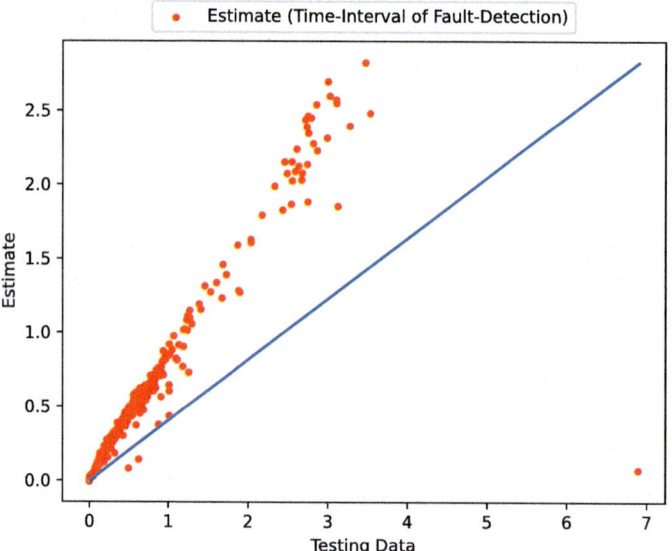

Fig. 10.9 The estimated scatter plot of time-interval of fault-detection without the case of jump diffusion process

10.3 Numerical Examples

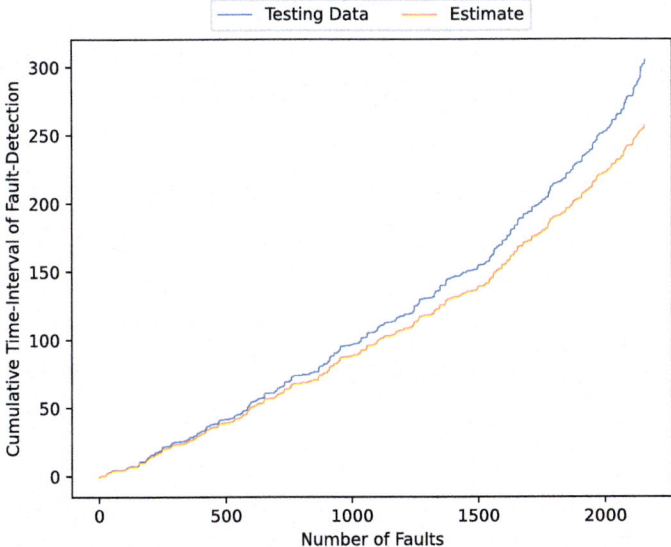

Fig. 10.10 The estimated cumulative time-interval of fault-detection without the case of jump diffusion process

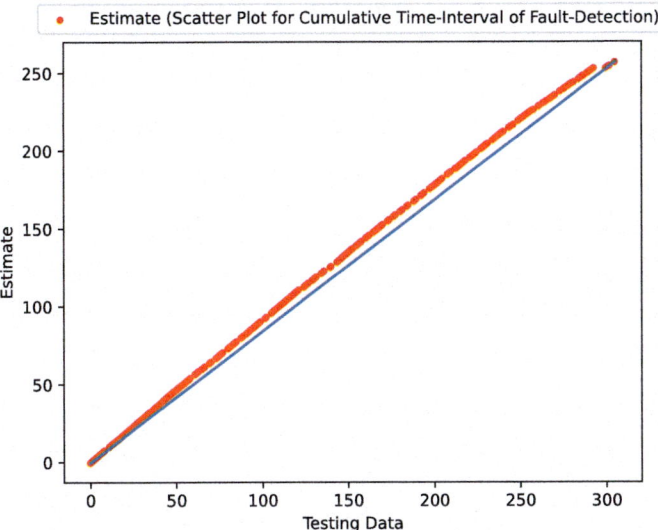

Fig. 10.11 The estimated scatter plot for the cumulative time-interval of fault-detection the case of jump diffusion process

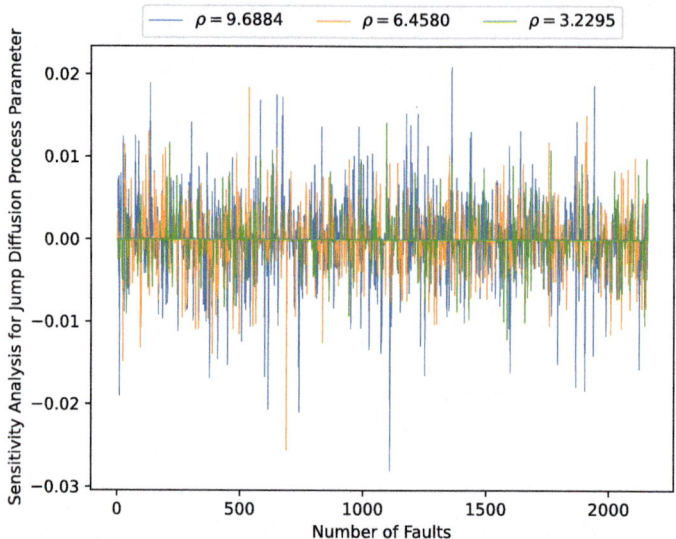

Fig. 10.12 The sensitivity analysis for the jump term changing the Poisson point process parameter ρ in the case of jump diffusion process

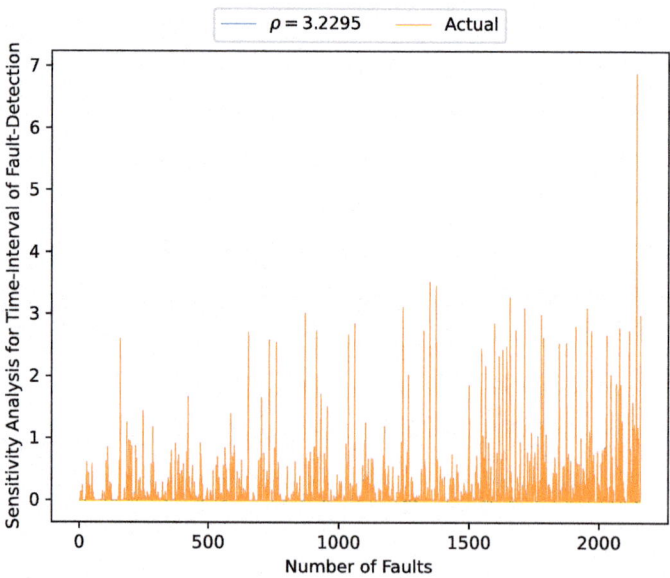

Fig. 10.13 The sensitivity analysis for the time-interval of fault-detection changing the Poisson point process parameter ρ in the case of jump diffusion process

10.3 Numerical Examples

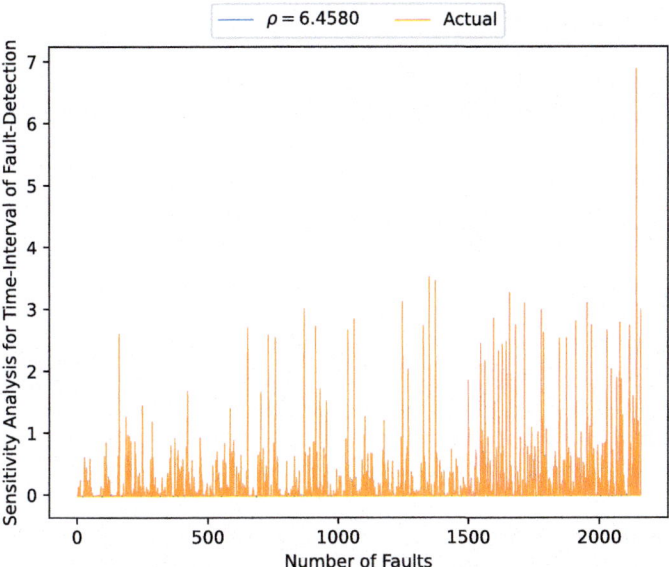

Fig. 10.14 The sensitivity analysis for the time-interval of fault-detection changing the Poisson point process parameter ρ in the case of jump diffusion process

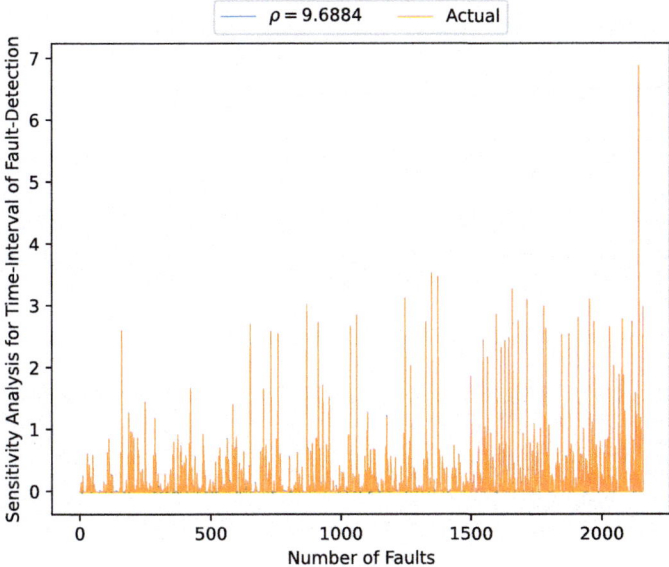

Fig. 10.15 The sensitivity analysis for the time-interval of fault-detection changing the Poisson point process parameter ρ in the case of jump diffusion process

10.4 Concluding Remarks

At present, many OSS's have been developed all over the world. This chapter has proposed the deep learning model assuming the unexpected irregular fluctuations as the characteristics of edge server and OSS. It is difficult for the edge computing managers to control the reliability under the cloud and edge computing. This chapter has discussed the deep learning based on the jump diffusion process as the characteristics of edge and OSS by using the actual edge OSS fault data as follows:

- The unexpected irregular fluctuation is considered as the environment of OSS fault-detection.
- The data preprocessing in the fault-detection process differ from the general data in the open data sets.
- The imperfect debugging process and version-upgrade is considered by using the jump diffusion process.

The proposed machine learning approach by deep learning will be helpful as the method of reliability assessment based on the time-interval of fault-detection under edge OSS service in operation phase.

References

1. Ibrahim IM et al (2018) A robust generic multi-authority attributes management system for cloud storage services. In: IEEE transactions on cloud computing. https://doi.org/10.1109/TCC.2018.2867871
2. Ahmad AA et al (2019) Scalability analysis comparisons of cloud-based software services. In: Journal of cloud computing: advances. Systems and applications. https://doi.org/10.1186/s13677-019-0134-y
3. Ozcan MO, Odaci F, Ari I (2019) Remote debugging for containerized applications in edge computing environments. In: Proceedings of the 2019 IEEE international conference on edge computing (EDGE), Milan, Italy, pp. 30–32. https://doi.org/10.1109/EDGE.2019.00021
4. Ngoko Y, Cérin C (2017) An edge computing platform for the detection of acoustic events. In: Proceedings of the 2017 IEEE international conference on edge computing (EDGE), Honolulu, HI, USA, pp. 240–243. https://doi.org/10.1109/IEEE.EDGE.2017.44
5. Caprolu M, Di Pietro R, Lombardi F, Raponi S (2019) Edge computing perspectives: architectures, technologies, and open security issues. In: Proceedings of the 2019 IEEE international conference on edge computing (EDGE), Milan, Italy, pp. 116–123. https://doi.org/10.1109/EDGE.2019.00035
6. Dolui K, Datta SK (2017) Comparison of edge computing implementations: Fog computing, cloudlet and mobile edge computing. In: Proceedings of the 2017 global internet of things summit (GIoTS), Geneva, Switzerland, pp. 1–6. https://doi.org/10.1109/GIOTS.2017.8016213
7. Yamada S, Tamura Y (2016) OSS reliability measurement and assessment. Springer series in reliability engineering. Springer
8. Yamada S (2014) Software reliability modeling: fundamentals and applications. Springer-Verlag, Tokyo/Heidelberg

9. Lyu MR (ed) (1996) Handbook of software reliability engineering. IEEE Computer Society Press, Los Alamitos, CA, U.S.A
10. Musa JD, Iannino A, Okumoto K (1987) Software reliability: measurement, prediction, application. McGraw-Hill, New York
11. Kapur PK, Pham H, Gupta A, Jha PC (2011) Software reliability assessment with OR applications. Springer-Verlag, London
12. Tamura Y, Watanabe H, Yamada S (2020) OSS project assessment based on discriminant analysis and jump diffusion process model for fault big data. Am J Oper Res 10(6):269–283
13. Tamura Y, Sone H, Yamada S (2020) Flexible jump diffusion process models for open source project with application to the optimal maintenance problem. In: International journal of reliability, quality and safety engineering, vol. 27, No. 6, World Scientific, pp. 2050020-1–2050020-18
14. Tamura Y, Yamada S (2020) Project maintenance effort optimization based on flexible JDP model for OSS fault big data. Int J Math Eng Manage Sci 5(1):66–75
15. Tamura Y, Sone H, Yamada S (2019) Productivity assessment based on jump diffusion model considering the effort management for OSS project. In: International journal of reliability, quality and safety engineering, vol. 26, No. 5, World Scientific, pp. 1950022-1–1950022-22
16. Tamura Y, Yamada S (2019) Maintenance effort management based on double jump diffusion model for OSS project. In: Annals of operations research. https://doi.org/10.1007/s10479-019-03170-w, Springer US, Online First, pp. 1–16
17. Tamura Y, Yamada S (2015) Reliability analysis based on a jump diffusion model with two Wiener processes for cloud computing with big data. In: Entropy, vol. 17, No. 7, Multidisciplinary Digital Publishing Institute, Switzerland, pp. 4533–4546
18. Merton RC (1976) Option pricing when underlying stock returns are discontinuous. J Finan Econ 3(1–2):125–144
19. The OpenStack project, Build the future of Open Infrastructure, https://www.openstack.org/

Performance Illustrations of the Developed Application Tool Based on Deep Learning

11.1 Data Set for Edge OSS Computing

We focus on the *OpenStack* Project [1] which includes several edge components.

In this chapter, we show numerical examples by using data sets on the assumption of the edge OSS service. The data used in this chapter are collected from the bug tracking system *OpenStack* Project [1].

The demonstration of our prototype tool is available from "DEMO APPLICATION" at the following URL; however, the function of calculation cannot execute considering the security:
http://www.tam.eee.yamaguchi-u.ac.jp/, accessed on 23 December 2023.

Our prototype tool has been released as the OSS based on GNU General Public License (GPL) in December 2023. The source code of our tool is available from "SOFTWARE" at the following URL:
http://www.tam.eee.yamaguchi-u.ac.jp/, accessed on 23 December 2023.

The total number of lines of data is about 20,000 lines. Then, the data consist of about 140,000 data items total. These are the specified version data. Actually, the users can obtain a greater number of data according to various OSS projects.

Our tool has been developed in Chap. 8.

11.2 Estimation Results

We analyze the fault big data in terms of fault-detection time in the OSS component of edge computing included under cloud computing such as OpenStack [1]. Table 8.2 is a part of data set. All lines of data is over 20,000 lines. Then, the all data is over 260,000 data items. Actually, the users can obtain a greater number of data according to various OSS projects. In this chapter, we show numerical examples by using a part of data set as shown in Table 8.3 on the assumption of the edge OSS service. We can obtain the fault-detection times from the "High" factors of bug levels in the bug-tracking system. In this chapter, we discuss the fault-detection time of the high fault severity levels such as "High".

First, we show the main screen of our tool in Fig. 11.1. In addition, Fig. 11.2 shows the menu of our tool. Moreover, Our tool is structured by using the dynamic link based on NW.js and Python. From the above-mentioned characteristics, we show the structure of our tool by using the package diagram. Then, Fig. 11.3 shows the package diagram of the prototype developed by using UML.

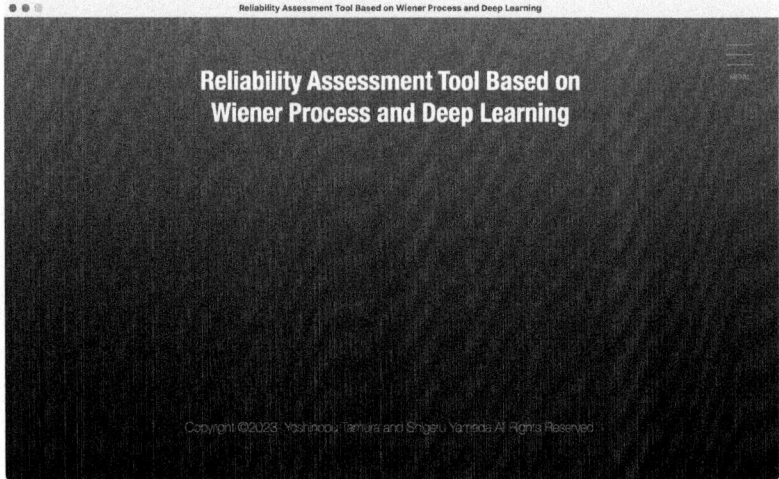

Fig. 11.1 The main screen of our tool

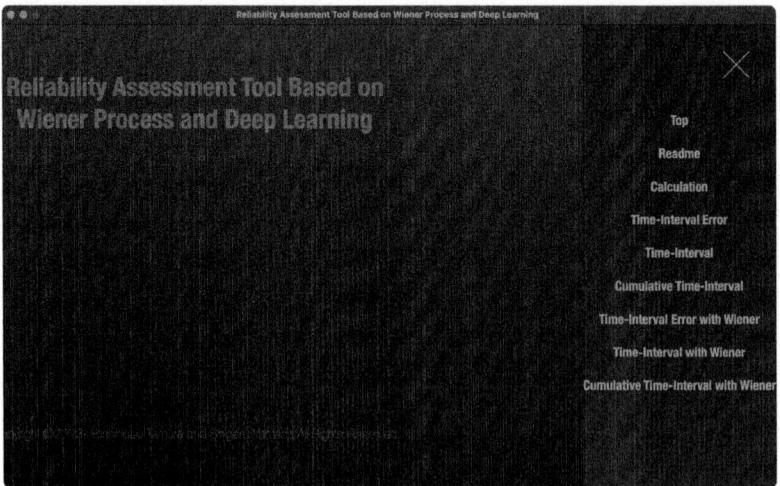

Fig. 11.2 The readme and menu screens of our tool

11.2 Estimation Results

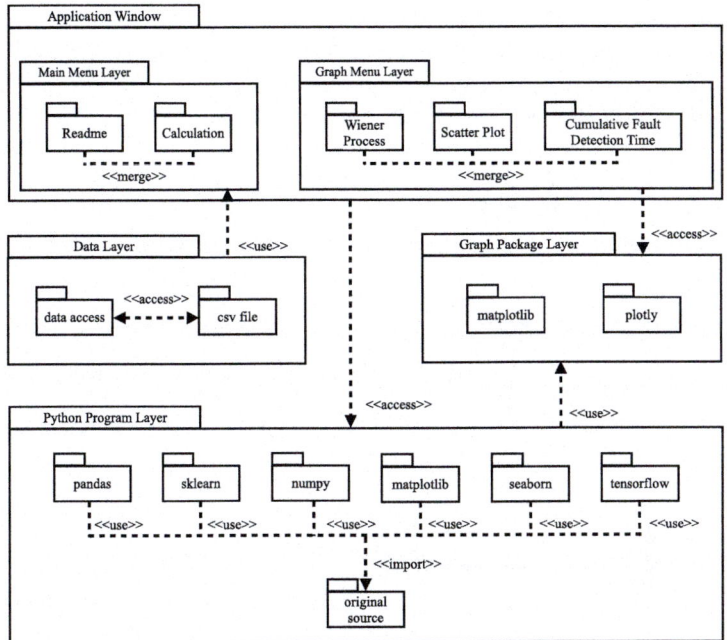

Fig. 11.3 The package diagram

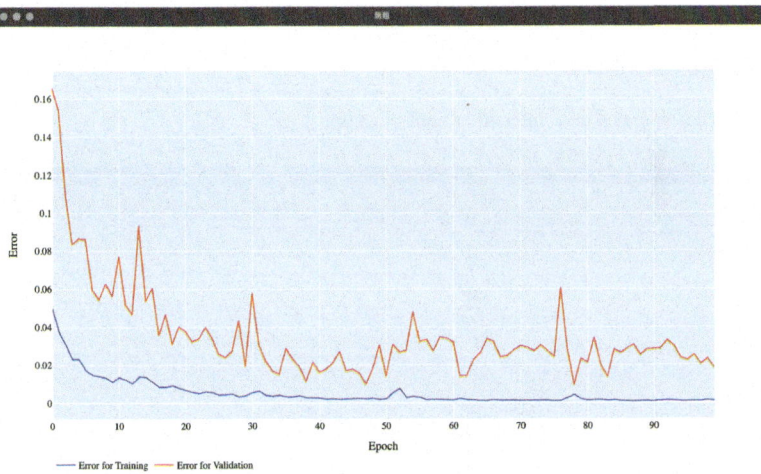

Fig. 11.4 The errors for training and validation

In the case without Wiener process, Fig. 11.4 show the overall pictures of the estimated errors between validation and training in the case of 10% testing data. From Fig. 11.4, we find that the errors of validation and training fit better in the case of 10%. In particular, the error of the "High" class fits better. Figures 11.5 and

11.6 show the estimated time-interval of fault-detection, the estimated cumulative time-interval of fault-detection. From Figs. 11.5 and 11.6, we find that the estimation results fit better until the 1000 faults.

In the case of Wiener process, Fig. 11.7 show the overall pictures of the estimated errors between validation and training in the case of 10% testing data. From Fig. 11.7, we find that the errors of validation and training fit better in the case of 10%. In particular, the error of the "High" class fits better. Figures 11.8 and 11.9 show the estimated time-interval of fault-detection, the estimated cumulative time-interval of

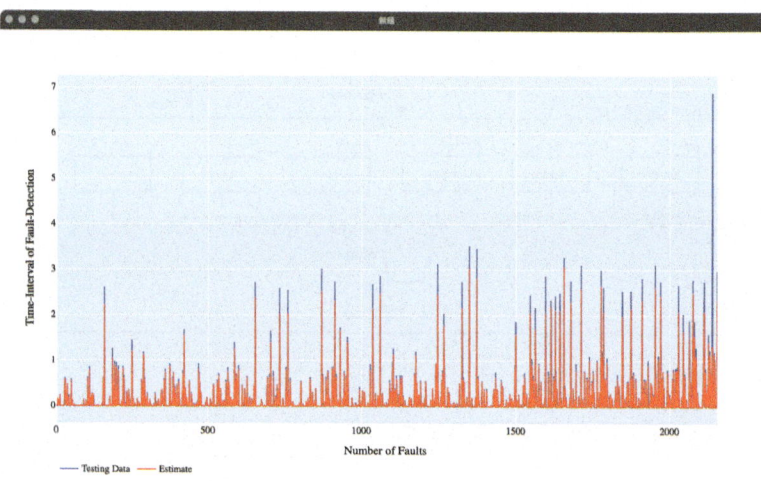

Fig. 11.5 The estimated time-interval of fault-detection

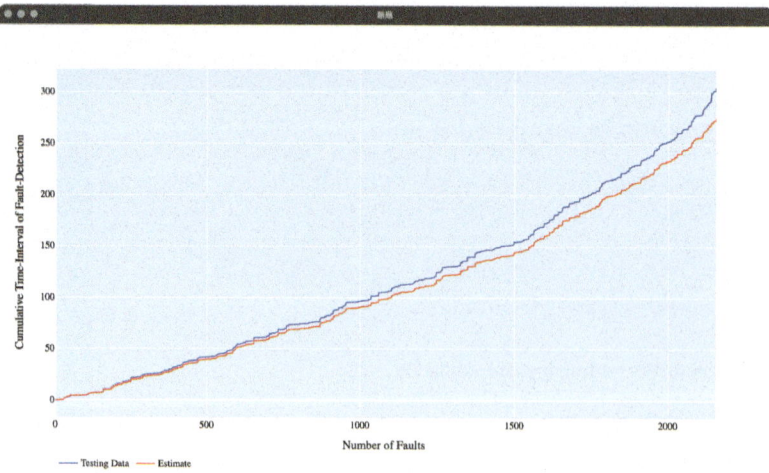

Fig. 11.6 The estimated cumulative time-interval of fault-detection

11.2 Estimation Results

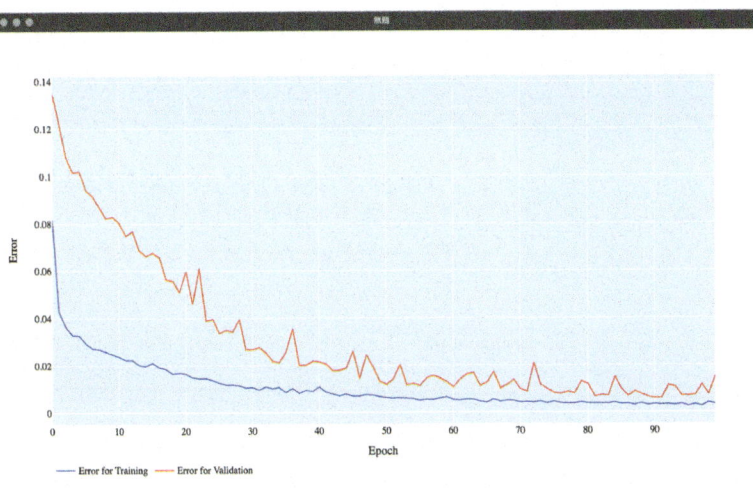

Fig. 11.7 The errors for training and validation with Wiener process

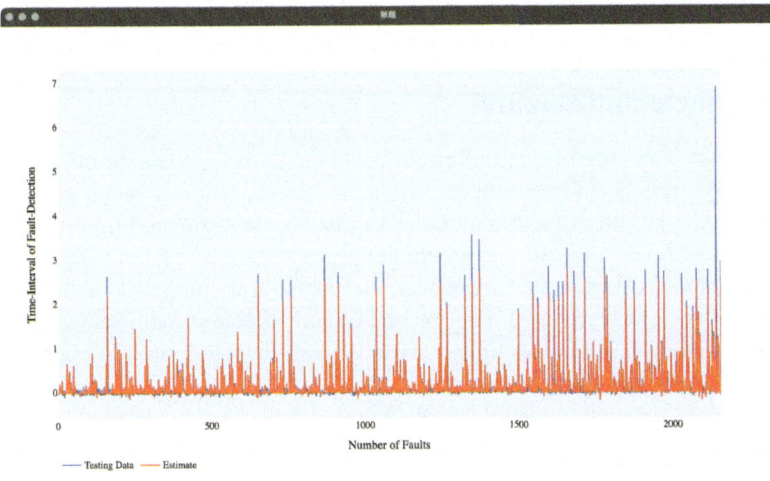

Fig. 11.8 The estimated time-interval of fault-detection with Wiener process

fault-detection. From Figs. 11.8 and 11.9, we find that the estimation results fit better all operation faults.

As for the above-mentioned results, we confirm that the developed prototype tool for reliability assessment based on deep learning with Wiener process is useful for estimating reliability in the near future. In particular, the advantage of our method is that it can make use of all the data on the bug tracking system.

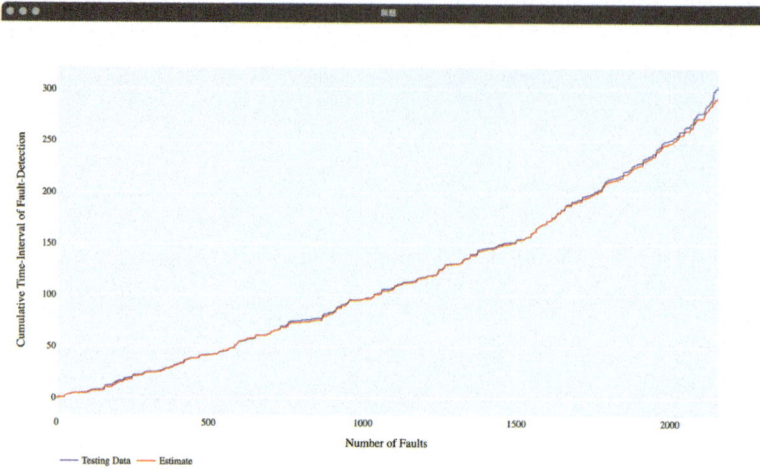

Fig. 11.9 The estimated cumulative time-interval of fault-detection with Wiener process

11.3 Concluding Remarks

This chapter shows performance illustrations of our tools based on the deep learning with Wiener process. For examples, the main screen, readme, menu screens are shown. Also, the errors for training and validation, the estimated time-interval of fault-detection, and the estimated cumulative time-interval of fault-detection are shown as numerical examples. Moreover, the errors for training and validation with Wiener process, the estimated time-interval of fault-detection with Wiener process, and the estimated cumulative time-interval of fault-detection with Wiener process are shown as the comparison results.

Finally, our tool can be useful for the OSS developers and managers to assess the OSS reliability based on deep learning with Wiener process by using the fault big data on the bug tracking system.

Reference

1. The OpenStack project, Build the future of Open Infrastructure, https://www.openstack.org/

Exercise 12

This chapter shows several exercises for understanding the reliability assessment measures for OSS reliability measurement and assessment. The following problems will be useful for software managers to evaluate OSS quality/reliability.

12.1 Exercise 1

Figure 12.1 shows the estimated expected number of remaining faults. Discuss the estimated expected number of remaining faults at 8,000 days in Fig. 12.1.

Brief Solution

It is important for software managers to decide the optimum OSS version upgrade time. At 8,000 days from Fig. 12.1, we can confirm that the number of remaining faults is about 10,000 faults.

12.2 Exercise 2

Figure 12.2 shows the estimated total software cost. Discuss the optimum release time and the cost in Fig. 12.2.

Brief Solution

It is important for software managers to decide the optimum OSS version upgrade time. Figure 12.2 show that the optimum version upgrade time is about 14,000 days, and the cost is about 130,000.

© The Author(s), under exclusive license to Springer Nature Switzerland AG 2024
Y. Tamura and S. Yamada, *Applied OSS Reliability Assessment Modeling, AI and Tools*, Springer Series in Reliability Engineering,
https://doi.org/10.1007/978-3-031-64803-8_12

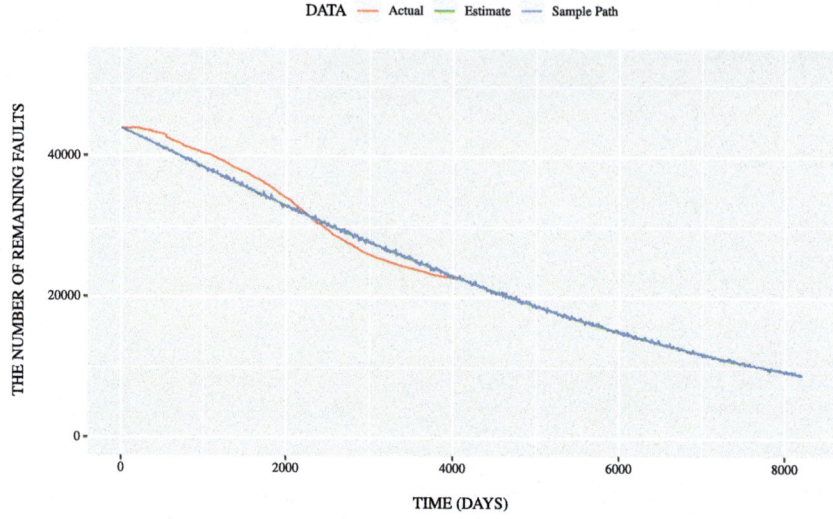

Fig. 12.1 The estimated expected number of remaining faults

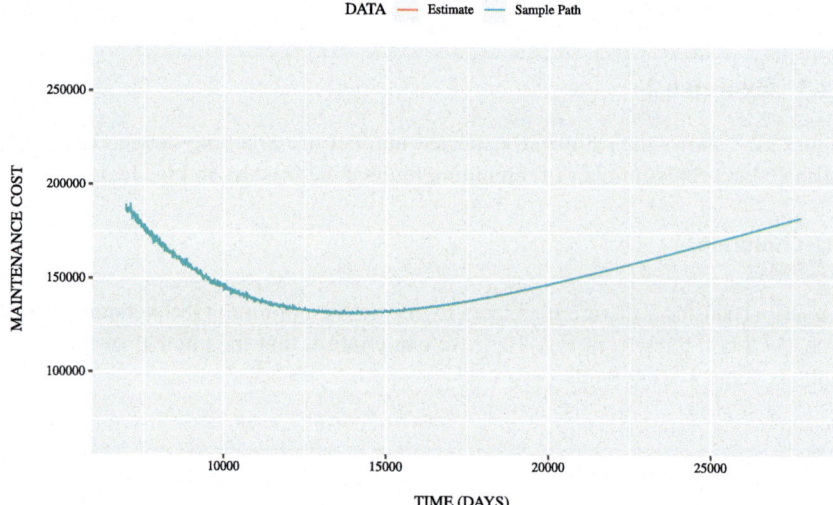

Fig. 12.2 The estimated total software cost

12.3 Exercise 3

Figure 12.3 shows the estimated instantaneous detection time of software faults by using the deep learning. Discuss the condition of reliability in Fig. 12.3.

Brief Solution

Figure 12.3 means that the OSS reliability is not growing, because the estimated instantaneous detection time of software faults becomes small according to the operation procedure is go on.

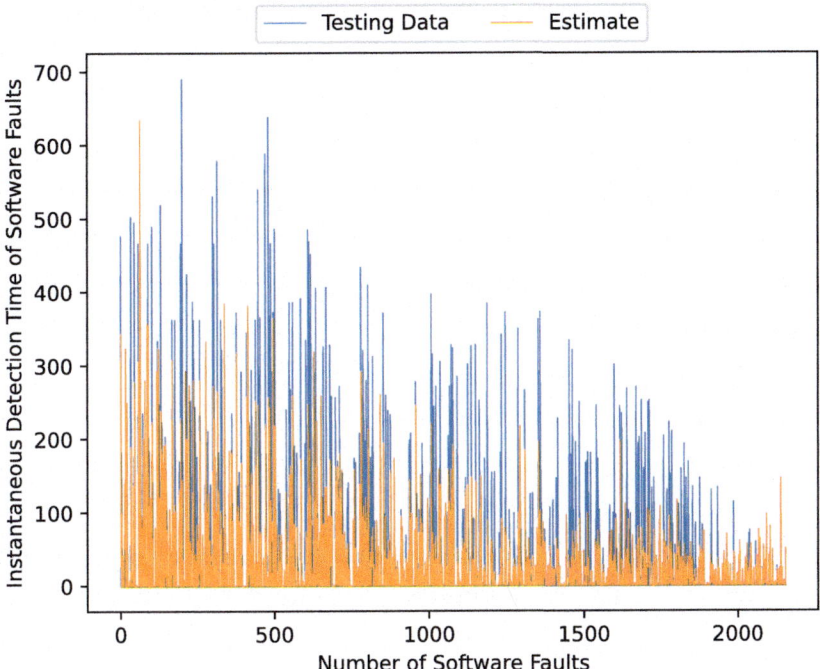

Fig. 12.3 The estimated instantaneous detection time of software faults

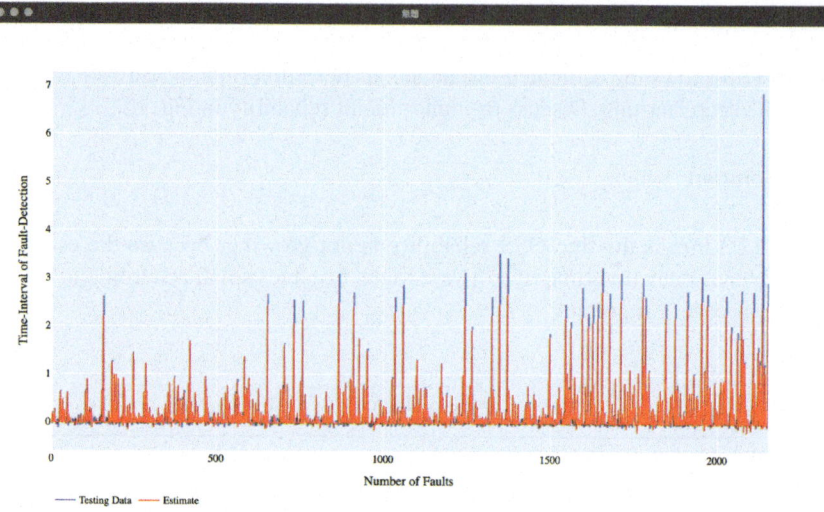

Fig. 12.4 The estimated time-interval of fault-detection with Wiener process

12.4 Exercise 4

Figure 12.4 shows the estimated time-interval of fault-detection with Wiener process. Discuss the condition of reliability in Fig. 12.4.

Brief Solution

Figure 12.4 means that the OSS reliability growth, because the estimated instantaneous detection time of software faults becomes large according to the operation procedure is go on.

SPRINGER NATURE

GPSR Compliance

The European Union's (EU) General Product Safety Regulation (GPSR) is a set of rules that requires consumer products to be safe and our obligations to ensure this.

If you have any concerns about our products, you can contact us on ProductSafety@springernature.com

In case Publisher is established outside the EU, the EU authorized representative is:

Springer Nature Customer Service Center GmbH
Europaplatz 3
69115 Heidelberg, Germany

The manufacturer's authorised representative in the EU is Springer Nature Customer Service Centre GmbH, Europaplatz 3, 69115 Heidelberg, Germany. If you have any concerns regarding our products, please contact ProductSafety@springernature.com

Printed and bound by CPI Group (UK) Ltd, Croydon, CR0 4YY

25/03/2026

02078169-0015